The Contemporary Internet

Participation in Broadband Society
Edited by Leopoldina Fortunati / Julian Gebhardt / Jane Vincent

Volume 3

PETER LANG
Frankfurt am Main · Berlin · Bern · Bruxelles · New York · Oxford · Wien

Leslie Haddon (Ed.)

The Contemporary Internet

National and Cross-National European Studies

PETER LANG
Internationaler Verlag der Wissenschaften

Bibliographic Information published by the Deutsche Nationalbibliothek
The Deutsche Nationalbibliothek lists this publication in the Deutsche Nationalbibliografie; detailed bibliographic data is available in the internet at http://dnb.d-nb.de.

ISSN 1867-044X
ISBN 978-3-631-60098-6
© Peter Lang GmbH
Internationaler Verlag der Wissenschaften
Frankfurt am Main 2011
All rights reserved.

All parts of this publication are protected by copyright. Any utilisation outside the strict limits of the copyright law, without the permission of the publisher, is forbidden and liable to prosecution. This applies in particular to reproductions, translations, microfilming, and storage and processing in electronic retrieval systems.

www.peterlang.de

Acknowledgements

Most of the work in this book originated in COST Action 298, in a sub-group called 'The multiple cultures of the European Information Society'. The exception is chapter two, which started out as a paper given at one of the COST 298 conferences[1]. Many thanks to Boldur Barbat, Bojan Srdjevič, Sandor Dembitz, Veljko Malbasa and Tomaž Turk for checking the draft chapters and helping prepare the final text.

The editor wishes to thank staff at Peter Lang GmbH, Berlin for their support in establishing a new series on new ICT and society called Participation in Broadband Society.

This publication is supported by COST and their staff are acknowledged for their assistance together with the COST Action 298 Chair Bartolomeo Sapio and Vice Chair Tomaž Turk. Thanks go also to the series editors for their sup-port during the production of this book.

COST - the acronym for European Cooperation in Science and Technology - is the oldest and widest European intergovernmental network for cooperation in research. Established by the Ministerial Conference in November 1971, COST is presently used by the scientific communities of 36 European countries to cooperate in common research projects supported by national funds.

"The funds provided by COST less than 1% of the total value of the projects support" the COST cooperation networks (COST Actions) through which, with EUR 30 million per year, more than 30 000 European scientists are involved in research having a total value which exceeds EUR 2 billion per year. This is the financial worth of the European added value which COST achieves.

---

1 The good, the bad and the unexpected: The user and the future of information and communication technologies, Institute of the Information Society, Moscow, Russian Federation 23-25 May 2007.

A "bottom up approach" (the initiative of launching a COST Action comes from the European scientists themselves), "à la carte participation" (only countries interested in the Action participate), "equality of access" (participation is open also to the scientific communities of countries not belonging to the European Union) and "flexible structure" (easy implementation and light management of the research initiatives) are the main characteristics of COST.

As precursor of advanced multidisciplinary research COST has a very important role for the realisation of the European Research Area (ERA) anticipating and complementing the activities of the Framework Programmes, constituting a "bridge" towards the scientific communities of emerging countries, increasing the mobility of researchers across Europe and fostering the establishment of "Networks of Excellence" in many key scientific domains such as: Biomedicine and Molecular Biosciences; Food and Agriculture; Forests, their Products and Services; Materials, Physical and Nanosciences; Chemistry and Molecular Sciences and Technologies; Earth System Science and Environmental Management; Information and Communication Technologies; Transport and Urban Development; Individuals, Societies, Cultures and Health. It covers basic and more applied research and also addresses issues of pre-normative nature or of societal importance.

Web: http://www.cost.eu

ESF provides the COST Office through an EC contract

COST is supported by the EU RTD Framework programme

COST 298 – Participation in the Broadband Society

Contents

Introduction: Leslie Haddon — 9

Part one: Cultural influences on the contemporary internet

Chapter one: Cultural factors shaping the experience of information and communication technologies. Frank Thomas and Leslie Haddon — 17

Chapter two: Young Italians' crossmedia cultures. Giovanna Mascheroni, Francesca Pasquali, Barbara Scifo, Anna Sfardini, Matteo Stefanelli, Nicoletta Vittadini — 33

Chapter three: Cultural influences on the adoption of web 2.0 services. Frank Thomas, Nicoletta Vittadini and Pedro Gómez-Fernández — 55

Chapter four: Cross-cultural differences in press coverage of the internet. Robert Pinter, Fruzsina Gyenes, Francesca Pasquali, Annika Bergström and Leslie Haddon — 77

Part two: Uneven experiences of the contemporary internet

Chapter five: Implications of the variation in broadband speeds over time. Leslie Haddon and Peter Heinzmann — 95

Chapter six: Cross-national broadband digital divides. Vesna Dolničar, Charalambos Christou, Rosemarie Gannon, Leslie Haddon, Soulla Louca, Pedro Puga and Jorge Vieira — 121

Chapter seven: eGovernment and the digital divide. Pedro Puga — 139

Chapter eight: The take-up of music 2.0. Jorge Vieira — 155

Part three: Methodological issues in comparative analysis

Chapter nine: Methodological issues in the cross-national analysis of contextual data. Leslie Haddon — 177

Chapter ten: Measuring the dynamics of cross-national digital divides. Vesna Dolničar — 191

Authors — 207

Leslie Haddon

Introduction

Overview

Any book dealing with how the internet is evolving could potentially address a vast range of topics. Obviously there is the question of how people are using it nowadays, an understandable theme of the majority of chapters in this volume. One could ask about how the governance of the internet is developing (i.e. the various ways it which it is regulated), or how its political economy is changing, by looking at the economic interests at work in the online world. There are potential questions about the nature of new types of virtual world to be found on the internet, the new spaces with related forms of communication and self-presentation, if one thinks, for example, of the growth of social networking sites over recent years (boyd, 2007). There is scope for thinking about the relation between online and offline experiences, in terms of the communities we belong to, how we socially network or how family members relate to each other. There are the questions about how much the internet continues to have implications for political engagement, as in the eDemocracy discussions (e.g. Bennet, 2008) or how it effects particular spheres of life, such as journalistic practices (e.g. Allan, 2006; Thorsen, 2008). And of course, one can consider a variety of ongoing (and new) issues – and hence policy interests – relating to the internet, ranging from copyright infringement in the distribution of audio-visual material online to cyberbullying (Smith, 2009)[1].

No one volume could deal with all of these elements, and as in many other publications the choices made in this book in large part reflect the particular interests of its contributors. That said, this book does cover a fairly broad range of developments, beyond looking at how people now use the internet to considering such questions as how its infrastructure is changing and how it is symbolically represented in the media.

When thinking specifically of the contemporary internet at the end of first decade of this century, the emergence of so-called 'web 2.0' applications has had a high visibility in the academic literature (e.g. Vossen & Hagemann, 2007). Although this may be an umbrella term covering very diverse services, as will be elaborated in subsequent chapters, these developments are associated with claims that they enable new forms of (more active) participation in the online world. This topic is explored at several points in this book, in chapter two

---

[1] For a general overview of perceived risks for children, see Livingstone and Haddon, 2009.

looking at the current practices of Italian youth, in chapter three's examination of factors shaping the take up of web 2.0 applications, and in chapter eight's focus on one particular web 2.0 development: music 2.0. But as indicated above, these types of usage questions, how we consume, communicate or participate online, are only part of the picture. Hence chapter five looks at how the broadband infrastructure has evolved in recent years, and how we therefore experience changing internet speeds, while chapter four examines press representations of the internet, suggesting ways in which they have become routinised in the news.

If there is one theme that re-appears across chapters it is the current manifestations of older digital divide issues. Addressing these, the second section of this book in particular examines the sometimes very uneven experiences of the internet's possibilities within and across countries. There is a major debate about the consequences of such differences, but the terms of this debate were first set up when most people still had narrowband access, and consequently used a more restricted range of applications. Therefore, one question that pervades several chapters is how the digital divide is evolving in the light of the more contemporary developments outlined above, be that in terms of uneven broadband speeds more generally (chapter five), practices relating to narrowband and broadband (chapter six), access to eGovernment services (chapter seven) or uneven participation in the world of music 2.0 (chapter eight).

Turning now to the comparative cross-national dimension of this volume, our interest in 'cultural' dimensions is a central theme, covered especially in the first section of the book. Although this is a problematic concept whose boundaries are unclear, potential cultural elements are identified when setting the scene in chapter one and these are subsequently operationalised in the research comparing survey data in chapter three. Once again, the cultural aspects are made manifest in a variety of ways, in terms of media representations in chapter four and in the case study of music 2.0 in chapter eight.

As regards the comparative element noted in the title, the final methodological section of the book illustrates some of the research challenges in this field. For example, in chapter nine we see the issues arising once comparative studies with qualitative elements start to involve many countries. If we want to compare those countries this leads to a pressure to impose some order upon them, such as grouping them according to an underlying criterion – which in effect is converting them into a quantitative measure. Meanwhile, chapter ten shows how quantitative studies face their own particular challenges. We are reminded that we can represent patterns differently using different statistics (and different graphs) and these representations then have a bearing upon how we interpret those data.

Beyond these two chapters, many of the others show the different methodological strategies used to maintain this comparative agenda. These range from the strategies involved in comparisons involving many countries (chapter three), those involving just a few countries (chapters four, five and six) or those involving just two countries (chapter seven). Where certain comparative data are simply not available, the chapter authors provide at least some speculation about what country differences we might consider (chapter eight) or what further comparative research is needed (chapter seven). Finally, even the single-country Italian study (chapter two) reflects on some of the specificities of the country concerned.

Chapter outlines

Cultural influences on the contemporary internet

Chapter one: Cultural factors shaping the experience of information and communication technologies

As preparation for the other chapters in this section, chapter one by Thomas and Haddon explores the diverse ways in which researchers have examined the effects of culture on the experience of information and communication technologies (ICTs). After some observations about the problematic nature of the term 'culture' and grey areas as regards whether to count some influences as 'cultural', the authors illustrate the breadth and depth of potential cultural influences (and cultural differences) relating to ICTs through contemporary and historical examples of research. This includes social structural factors (such as the degree of homogeneity versus heterogeneity within a society and gender roles), time structures, value systems, communication forms and material culture (as embodied, for example, in housing characteristics and artefacts used in daily life).

Chapter two: Young Italians' cross-media cultures

Some of the influences outlined above are illustrated in chapter two by Mascheroni et al, in an empirical study of Italian youth's changing ICT consumption in the face of an evolving media landscape. Drawing on the domestication framework, the authors document some of the key factors affecting young people's media choices, covering intra-family interactions, relations with peers, the symbolic meaning of media, financial considerations and the constraints imposed by the everyday time structures in which people operate. What emerges is a complex picture of the interrelationships between ICTs, where the authors explore 'cross media consumption' (for example, interests that originate from viewing TV lead to the collection of related internet material and newer practices such as the storage of audiovisual content for trading, sharing, gift-giving). Although a national study, it reflects on what might be more specific to the Italian context.

Chapter three: Cultural influences on the adoption of web 2.0 services

While the effect of cultural differences between countries on the adoption of internet services has been under-researched in general, in chapter three Thomas et al argue that these might be even more significant for web 2.0 services. The authors present an analysis of the differences in web 2.0 service adoption among European Countries. This is based upon a secondary analysis of data concerning the various cultural influences and social behaviour that can affect adoption. The results show how these factors can facilitate or constrain web 2.0 services usage, affecting different types of services in the diverse ways. The main difference between countries is in terms of online practices relating to what we might call 'cultural production' (e.g. sharing material or producing user generated content) and online social networking. Cultural production through web 2.0 services is more diffused in countries where there is generally more participation in public life. In contrast, social networking online is more diffused in countries where, overall, people are engaged in fewer social activities offline. In the first case, the diffusion of particular web 2.0 services seems to follow from pre-existing ways of participating in social life whereas in the second case online forms of social networking appear to be adopted more where they offer new opportunities for social participation.

Chapter four: Cross-cultural differences in press coverage of the internet

Media representations of the internet can influence people's perception of how much the online world is leading to changes in their lives, their evaluation of whether this is for better or worse, and their understanding of what is happening online and what they can do there. While we now have various tools for analysing how and why the media organise coverage and with what implications, there is little research specifically on media coverage of the internet and little, also, on cross-national media reporting – certainly in the field of ICTs. In chapter four Pinter at al draw on original empirical research, a content analysis of press coverage of the internet in four countries, in order to demonstrate the various levels on which that the portrayal of the internet varies nationally and to suggest some of the processes that produce these different representations.

Uneven experiences of the contemporary internet

Chapter five: Implications of the variation in broadband speeds over time

While it is clear that internet speeds have increased over the years, applications have also evolved, including in terms of their size and complexity. So how have these changes affected the time taken to achieve different goals online, a consideration that affects both their feasibility but also their attractiveness to users? In chapter five Haddon and Heinzmann show how faster speeds have indeed had a bearing upon the time taken to use variety of applications, and how much and what kind of difference changes in speed have made to people's online experiences. They then proceed to indicate the degree to which the speeds advertised by ISPs differ from the actual speeds that different users can expect to experience in practice, because of delays in decisions to upgrade, distance from the local provider and congestion at different times of days. Lastly a cross-national component of the analysis demonstrates the variation in internet speeds over time between European countries, and, once again, indicates how these would be experienced.

Chapter six: Cross-national broadband digital divides

Chapter six by Dolničar et al draws upon insights from the digital divide debates in order to understand the differences resulting from the transition from narrowband to broadband internet access. In particular, through a secondary analysis of Eurostat data, the chapter contrasts the experience of Cyprus, Ireland Portugal, Slovenia and the UK to show the cross-national factors that might also be at work in this process. Much attention has been given to web 2.0 applications, and, tentatively, the authors suggest that broadband does facilitate some of these. But they also show how it supports older, more established online practices, which are themselves evolving (e.g. web page creation and down-loading software). However, the cross-national analysis shows how the popularity of different practices varies by county, suggesting that there are cultural or social factors are work as well as the technical possibilities offered by broadband.

Chapter seven: eGovernment and the digital divide

Usually the topics of eGovernment and the digital divide are not addressed in the same body of literature, but as many governments are moving towards a relation with their citizen that is increasingly mediated by the internet one can ask if this contributes to creating new forms of digital divide. In chapter seven Puga draws on Portuguese data to indicate both the range of government services available and planned, providing a first broad indication of their usage. To avoid being carried away by the opportunities associated with eGovernment, the author provides examples both of ways in which online services can give advantage to some compared to others, but also cases showing how the online versions of Government services can have their own limitations. In order to underline this complexity, the author discusses a variety of scenarios to illustrate where the inability to use eGovernment can be more or less of a problem, indicating the types of research we would need to conduct in order to evaluate the consequences of these services. Comparing data from the UK and Portugal reveals a common socio-demographic pattern as regards who uses different service. Finally, the chapter considers more recent eGovernment initiatives related to web 2.0 developments, like Second Life, and the ongoing scope for further social exclusion.

Chapter eight: The take-up of music 2.0

Music 2.0, involving the downloading and sharing of digitised music, is regarded by some as representing a paradigm shift in music consumption. In chapter eight Vieira first reviews the history of innovation in the field of music reproduction. This indicates the extent to which music 2.0 is new but also the extent to which it is part of a continuous evolution of consumption practices. The main part of the chapter then draws upon the measures developed within the digital divide literature – around the nuances of access, multiple measures of usage and skills and the nature of participation in the online world – in order to show the contours of music 2.0 adoption, illustrating this with Portuguese data. The author ends by considering potential measures of the significance of music 2.0 in people's live and explains what issues may be of interest for cross-national research on this topic.

Methodological issues in comparative analysis

Chapter nine: Methodological issues in the cross-national analysis of contextual data

Chapter nine considers some of the issues involved in the data analysis stage of a cross-national study of children's experience of the internet. Taking the project *EU Kids Online* as a case study, Haddon focuses on the challenges faced and the basis of decisions taken when analysing the influence of contextual factors within the countries that were studied. Specifically, the author considers the choices faced when comparing many countries and using a system of analysing national reports from the participating research teams. Three areas are considered: the choices regarding conducting quantitative verses qualitative analysis of the data, counting issues and the process of clustering counties. The overall aim is to share methodological insights from the experience of conducting such a study and to illustrate the types of strategies and forms of analysis that are possible.

Chapter ten: Measuring the dynamics of cross-national digital divides

The other methodological contribution to this volume, chapter ten, examines an innovatory approach to addressing the problem of how to make comparisons of the digital divide over time. Judging whether the digital divide is increasing, decreasing or constant even within one country turns out to be less than straightforward, depending very much on the particular measures examined. Dolničar outlines the nature and implications of various models of diffusion, discussing the relationships between different measures of the digital divide as a step towards developing a broader methodological framework in this field. Using Slovenian data, the author illustrate even more complexity in the dynamics of the digital divide, as it increases or decreases at different time points according to the various measures employed and the varied scenarios being used. The aim is show the principles of this methodological framework thorough a worked example.

Bibliography

Allan, S., 2006. *Online news: journalism and the internet.* Maidenhead: Open University Press.
Bennet L. W., 2008. Civic life online: learning how digital media can engage youth. Cambridge, MA: The MIT Press
boyd, D., 2007. Why youth ♥ social network sites. In: D. Buckingham, ed. Youth, identity and digital media. Cambridge, MA: MIT Press, pp. 119-142
Green, L., (2010) The internet. An introduction to new media. Oxford: Berg.
Livingstone, S. & Haddon, L., eds. 2009. Kids online. Opportunities and risks for children. Bristol: Policy Press.
Smith, S., 2009. Cyberbullying: Learning form the past, looking to the future. COST IS0801 workshop Cyberbullying: Definition and Measurement. Mykolas Romeris University, Vilnius, Lithuania 22-23 August 2009.
Thorsen, E., 2008. Journalistic objectivity redefined? Wikinews and the neutral point of view. New Media & Society 10 (6): 935-954.
Vossen, G. & Hagemann, S., 2007. Unleashing Web 2.0: From concepts to creativity. Boston: Elsevier/Morgan Kaufmann.

Frank Thomas and Leslie Haddon

Chapter one. Cultural factors shaping the experience of information and communication technologies

Introduction

This chapter[1] explores the diverse ways in which researchers have examined the effects of culture on the experience of information and communication technologies (ICTs). By illustrating this range of claims through contemporary and historical research examples, it aims to sensitise those involved in researching, using, designing or developing policy on ICTs to the breadth and depth of potential cultural influences and cultural differences. In other words, the chapter is not meant to be a theoretical statement on the nature of culture so much as a guide to what has in practice been considered by those studying ICTs.

We start with a working definition that we could use to explore in more detail the range of elements that can be considered to be 'cultural' without ruling out the possibility that some people could argue for an even more inclusive definition or alternative understandings of the term (e.g. as 'discursive action', Hester & Eglin, 1997). Therefore, in this chapter we shall understand by 'culture' some kind of commonly shared symbols, values, beliefs, and attitudes, as well as their translation into everyday social perceptions, behaviour and material artefacts.

The literature referring to culture indicates that it can exist at various levels, in a variety of forms. In principle, it could stretch from what have been identified by some as lifestyle subcultures (e.g. Hippies) through the virtual group culture of internet communities (e.g. Internet Relay Chat or 'IRC') to the business culture of a corporation. In practice, in this chapter we will focus mainly on cultural factors that have been associated with nation states, but at times indicate different conceptions of 'culture' within societies.

The problematic boundaries of culture

First, some caveats. 'Culture' is probably one of the most contested words within the social sciences. There are different definitions in sociology, cultural anthropology, cultural studies, media studies and social psychology. Hence, different disciplinary or professional backgrounds will make different readers

---

1   A more expanded review can be found in Thomas, et al. (2005)

sensitive to various claims about what counts as 'cultural' or not. Moreover, there will always be grey areas.

Let us consider some examples. Within Europe, the early launch of a common mobile phone standard in the Nordic countries in part explains why the pattern of take-up over time has been higher here than in some other European countries. This standardisation process involved decisions by the state-run PTTs[2] as well as by regulators. So would this count as an example of an influence that was not in some sense cultural? After all, it was, in part, an administrative decision. At one level this is true, but the staff involved in those early negotiations pointed to a tradition, at least in more recent times, of Nordic collaboration (COST248 Mobile Group, 1997). Would that then count as being cultural?

Another example might be the area of education. To what extent do the particular education systems and particular educational arrangements (like the timing of the school day) in different countries reflect cultural values, versus to what extent do they reflect historical, political and administrative decisions that could have been otherwise? Or when the state, or any other body, intervenes to ban or regulate the use of mobile phones in certain public spaces, how much does that reflect local cultural values and how much is it just a decision of the body concerned, perhaps reacting to media concerns or to particular lobbying interests at the time? Again, could that policy have been otherwise? To what extent was it contingent?

The second caveat to mention is the relationship between culture and spatial communities. Due to the long history of nation states in Western Europe the effects of culture can be confounded with those of a country. Yet, even in long-established nation states different communities have evolved side by side. For example, in multi-faith countries, such as in Switzerland or Germany, there are important differences within the state related to religious/ethnic identities and potentially values and orientations – which we might consider to be cultural. Meanwhile, language can provide a basis for communities that may be smaller than a country – e.g. the Basque language in Spain and France or the German, French or Italian speaking cantons in Switzerland[3]. Alternatively, language 'communities' – perhaps with some shared cultural aspects - can be larger than specific countries: such as French-, English-, German-, Dutch- and Swedish-speaking peoples, to name but a few. In short, culture should be treated as something different from, but may be related to, the influences of country, ethnicity, religion and language.

---

2   'Post, Telegraphs and Telephones'. The Post Office in various countries was also responsible at that time for telecommunications.
3   For example, Gilligan and Heinzmann (2004) demonstrate that differences between these communities exist in terms of TV watching and radio listening.

The third caveat, that takes the division of culture further within nation states, is that we may also choose to talk about the cultures of generations (e.g. the baby-boomers, generation x), of classes, more narrowly of professions, or of sub-cultures related to lifestyle, etc. – a range of other possibilities was noted earlier. Although this will not be so developed in this particular chapter, it is worth noting that these are other ways of looking at culture within and across countries.

Cultural Influences

Social structural factors

When we start to examine the different elements that have been considered in research on ICTs, there is a variety of what might be called social structural factors, such as the degree of homogeneity versus heterogeneity within a society. For example, a socially more homogeneous society – i.e. with largely shared common symbols, values, behaviours, language and institutions influenced by a dominant faith or political ideology – may well facilitate the diffusion of ICTs. The question is, and a question that we can ask for the following elements also, is whether some of these elements should be considered to be 'cultural' in nature? Certainly, on a smaller scale, diffusion studies have amply shown that an innovation spreads more easily in socially homogeneous than in heterogeneous social networks, provided that the social values of the network members are compatible with the use of that specific innovation (Rogers, 1995). For example, after the expiry of the Bell telephone patents at the start of the last century, telephone cooperatives organised the socially homogeneous networks of small marketing towns and their rural hinterlands states in Midwest U.S. Their networks resulted in an impressive telephone density that was actually higher for a time in the rural U.S than in its more urbanised counterparts (Fischer, 1987).

The hierarchical structure of a society can both facilitate and retard the development of communication media. For instance, the evolution of nation-states and empires and their hierarchical princely administrations in Renaissance Europe was paralleled by the construction of the first postal services to coordinate these bodies (Sautter, 1951). With the advent of colonial empires in the 19th century these networks were extended by submarine telegraph cable networks all over the world partly for the same reasons (Headrick, 1981). However, the spatial diffusion of the telephone in late 19th century France shows how a hierarchical social structure can also hinder the development of a communication medium. In France, rural areas were dominated by clientilistic power networks that used their position as intermediaries in a hierarchical communication structure between the rural regions and the capital, Paris, to control local power. The new technology of telephony threatened to undercut these information filters by enabling uncontrolled communications between rural departments and Parisian decision-makers. Therefore the local decision-makers who co-financed the telephone lines did not push themselves to help construct the new lines and so deliberately delayed the growth and the use of the French telephone (Carré, 1991). There are even more drastic examples of this negative effect on diffusion. For example the Sultan of the Ottoman Empire decided not to allow the establishment of a first telephone network in the capital Constantinople fearing that the new technology might undermine his autocratic rule. Meanwhile, Stalin halted the further residential diffusion of telephony in Soviet Russia while at the same time, the new Soviet Government established an up-to-date, all-Russian and centralised radio telegraphy network to command and control the bureaucracies of the party, the police and the state (Craemer and Franke, 1935).

Religion is one of the most important ingredients of a culture as it can influence the definition of the individual in society, especially the individual's sense of life, his or her liberty, the structure of communications (i.e. whether they may be more horizontally or more vertically oriented) and the development of an independent third organisational layer in society between the individual and the state (i.e. civil society). Religiously influenced values explain, in part, the position of women and the family and the structure of the educational, social welfare and health systems. So, the legacy of religious structures and values form basic matrices that can influence the ways that people communicate[4]. Illustrating this, research on the extent of 'interpersonal trust' (i.e. the degree to which an unknown person is trusted), examined in World Values Surveys, shows a drop in trust from Lutheran-influenced to mixed Lutheran-Catholic, to Catholic, Orthodox, and Muslim cultures. It remains to be seen whether this would translate into the different degrees of acceptance of media such as Internet Relay Chats or social networking sites, where chatters or visitors can meet complete strangers even if the site operator labels them as being 'friends'.

Education, both as a structural influence and as experienced by individuals can be considered to be partially cultural in nature. The structure of national educational systems, the stress laid on selecting the best students versus attaining more social inclusiveness, as well as the content of literacy programmes, are all related to such values and priorities. Several studies show that educational attainment, which varies both within and between countries is actually one of the major influences explaining levels of internet adoption and of drop-out rates (Rainie, 2003). This is also true for 'literacy', originally defined

---

4 There is also a far more direct influence of European monasteries: they established the first Trans-European letter services in the Middle Ages.

in terms of reading and writing but now including skill requirements in an information society under the term of 'digital literacy' (SIBIS, 2003). The mental capacity to manage abstract thinking, in part influenced by education, is very unevenly distributed in social terms and is a strong factor shaping ICT adoption and the successful integration of ICTs into the routines of everyday life (Weiss, 2001; Iske, et al., 2004).

As argued in chapter four, Media Studies research, alongside that of other disciplines, has highlighted ways in which the mass media affect our perceptions of reality, including perceptions of ICTs, from symbolically contributing to their fashion status to raising concerns about their social consequences. This can affect motivations to acquire these technologies, how they are used, and indeed how that use is regulated, if we think about parents making rules about children's use, for example. Of course, national mass media may themselves be influenced by the wider national culture. For instance, the *EU Kids Online* study discussed in chapters four and nine noted that the fact that pornography received less attention in the Norwegian press compared to some other national media may well reflect the widespread perception of child sexuality as being more natural in that country (Haddon & Stald, 2009). But of relevance to this chapter, the mass media may also be considered to be cultural influences as institutions, i.e. they have their own cultures. In this respect Hallin and Mancini (2004) discuss the different media system that correlate with different regions in Europe, whose countries are historically and culturally related. Meanwhile chapter four of this volume uses empirical data to show the different national media styles that influence how content is presented – from how it is classified, to what categories of news receive more attention to variations in whose voices are heard in the national media.

Ethnicity can play a role at both the collective and individual levels. Minority ethnic groups within countries can be organised so that communications pass through family and association channels. Jewish diaspora networks and the transnational communities of Dominicans in the U.S. provide examples of how easily ethnic communications are able to overcome the barriers of distance (Portes, 1997). On the individual level, ethnic background has played an important role in the diffusion of the internet in the US, as Afro-Americans and Hispanics have consistently lagged behind whites and Asians (Hoffman and Novak, 1998). Ethnographic research on West Indian immigrant communities' use of ICTs in the UK (Miller & Slater, 2000) as well as of immigrants' communications in the Netherlands and in France (Calogirou & Andren, 1997) all show convincingly that ethnic background strongly influences the intensity, the social composition and the geographical reach of social networks and how they are maintained over distance by use of the telephone, mobile phone or e-mail (LeRay, 1994).

Language is a major carrier of culture. The initial domination of the Internet by English-language websites is a well acknowledged issue for non-English speakers (Vehovar, et al., 1999; NTIA, 2000; Lazarus & Mora, 2000). Although there are of course web-sites and services available in a variety of languages, overall there is simply less content than for those who cannot speak English, even after the arrival and dramatic growth of the Chinese language on the web. A related point is actually noted in chapter eight regarding the more limited music 2.0 options open to those who do not speak English. More generally, the issue of language on the internet can be more of a barrier for non-English speaking older people (Gilligan, et al., 1998) and the less well-educated who are less likely to speak a foreign language. The influence of language can work at other, more subtle, levels as well as just being a barrier to the take-up of ICTs. For example, software is often supplied in English first and then other languages, affecting the timing of when it becomes more accessible to different language speakers. Lastly, it is also important to take into account the role of orality in a culture in relation to writing, the type of alphabet, the use of images in communication etc. Such factors may influence people's competencies concerning, for example, the use of i-Mode (notably the way the Japanese abbreviate messages).

Lastly, one could argue that structural factors could include various elements from what might be called the 'social constructionist' tradition of analysis[5], pointing to the ways in which expectations and understandings of roles are social constructed. For example, in what ways are gender roles experienced differently in different cultures, including the degree to which strict gender divisions are maintained across different aspects of life? To take a particular example, women's participation in the labour force varies, both in terms of the proportion of women working and the nature of that work, which can have a bearing upon personal disposable income and thus the capacity of women to acquire ICTs.

In the same spirit, how are children's roles (and parents' roles) socially constructed and experienced differently in various countries? This has potential implications for parent-children relationships around ICT (Haddon, 2004). For example, one pan-European study revealed different parenting styles in different European countries, as illustrated by the ways in which parents regulated their children's TV viewing (Pasquier, 2001).

Temporal structures

The time structures of different nations, but also of different social groups, can vary. Examples would be the timing of when work starts and ends (as well as the

---

5 For example, the social construction of childhood, see James and Prout, 1997.

length of the working day), the degree to which people engage in organised leisure activities, be that after school or after work, and differences in the timing of activities even in matters such as when people eat. Quite simply, these can all affect the timing of when people use ICTs, be that watching media, going on-line or communicating.

One study of the use of i-Mode's successful take-up in Japan evoked, amongst other things, an argument about time distribution (Heres, et al., 2004). This noted that Japan is a more outdoor-oriented society than many European countries, given that Japanese homes are small and lack privacy. Therefore people spend a good deal of time outside the home, which means that home-based-ICTs are not so attractive. The study argued that i-Mode became popular in part, but only in part, because having the internet in the home was not so appealing but internet-like services were nevertheless desired.

Turning to the subjective experience of time, one qualitative study involving focus groups from 6 European countries noted that there were some systematic national differences with regard to how people articulate their subjective experience of time (Klamer, et al., 2000), an observation confirmed in multi-cultural settings by Levine (1997). While many participants acknowledged that they led busy lives, in some countries there was a great willingness to talk about this in terms of time pressure and stress[6], whereas in others participants talked more about the importance of being in control of their own life, of avoiding stress – but not saying they expressed stress[7].

This subjective dimension is relevant for the use of ICTs since whether people in different cultures perceive problems with time might have a bearing upon decisions to adopt technologies that offer solutions in terms of time-saving or, more commonly, allowing the more flexible use of time. And such perceptions might have a bearing upon people's willingness to invest their time in acquiring and learning to use ICTs.

Lastly, we have to consider various cultural expectations about time (Levine, 1997). A first example would be social time norms – e.g. norms about when and when not to communicate. Apart from norms about how to make calls, how to speak, there are also ones about how long to call and when to make certain calls. As long ago as 1903, Simmel (1976) had observed that the time stress in modern cities is considerable and that social norms of punctuality shape the rhythms of urban sociability. A second example would be that in different countries there sometimes seem to be different expectations about how rigid the boundaries should be between work time and free time. For example, when comparing the responses of US and Dutch focus groups, this willingness to blur home and work times was one of the differences between the two national groups (Mante, 2002).

---

6 For example, in Spain and Italy in this study.
7 For example, in Denmark in this study.

Once again, this could influence the timing of when ICTs are used for work and for non-work purposes. In fact, there has been discussion of countries' orientations to time, with monochronic and polychronic time cultures[8] (Hall, 1983). One study utilised this concept to explain Singaporean people's ambivalence about the use of the mobile phone to arrange meetings when it led, for instance, to a decline in punctuality (Chung & Lim, 2005).

Value systems

An obviously relevant value identified in the ICT literature is 'openness to technological innovation'. In other words, while potential future users of technologies can be characterised by their general social position, their position within social networks, etc. the degree to which they are exposed to new information coming from outside, and their receptiveness to these innovations, is also important. This is partly used as a rationale for ICT companies to test their products in some countries first, where this value is high – such as the UK, Hong Kong and Japan. Meanwhile, diffusion studies have analysed the effect of specific sets of values on the speed of the diffusion of innovations and on the social setting of the innovators (Rogers, 1995).

One particular value distinction that occurs in a number of guises is that between an orientation towards being individualistic or to the group, in whatever form. For example, one Italian study hypothesised that the mobile phone was so popular with Italians because of the individualism and the great flexibility that they have developed in the world of work (above all in regions of advanced capitalism, for example in North East Italy) (Fortunati, 1997). In this case, individualism is cited as a factor shaping the rate of adoption of an ICT. In contrast, one study of Korean life argued that people in that country are often considered to be members of families more than individuals (Yoon, 2002). This is illustrated in the way that young people do not really have personal space in the home (e.g. their rooms are accessible to other family members without permission), which can in turn influence the nature of ICT adoption. Moreover, ICTs (like the PC) are often familial rather than individual possessions. The study also argued that in relation to the mobile phone, the above values mean that for many young people, calls from parents are more significant than calls from peers, they are seen as a form of 'mobile affection', an expression of family bonding. Hence, the study argues that the specific orientation to family in Confucian Asia (i.e. also China and Japan) makes a difference to the use of the

---

8   To take one important dimension of these concepts, in monochronic cultures, people adhere more strictly to schedules while in polychronic ones they change plans more easily.

mobile phone compared to the Western studies that often stress how children use the mobile phone to be more independent of parents.

Finally, the work of Hofstede[9] (1980) and Trompenaars (1993), looking at cultural norms originally from a managerial perspective, has been cited in some studies of ICTs, e.g. one looking at different patterns of adoption of the internet across countries (Thomas & Mante-Meijer, 2001). If we take the work of Trompenaars, these cultural norms include elements such as individualism vs. collectivism, whether cultures have universalist or particularist orientations, whether cultures are specific or diverse, whether cultures are affective or neutral, how cultures accord status (whether it is ascribed and achieved), and how cultures relate to nature (e.g. controlling it vs. letting it run its course). Sundqvist, Frank and Puumalainen (2005) found that uncertainty avoidance (i.e. the propensity to avoid risks and to follow established rules) influenced the speed of mobile telephony diffusion. The prevalence of the value of uncertainty avoidance reduced the adoption speed in early adopter countries, since accepting uncertainty is necessary for trying a new product, but then increased it in later adopting countries. This confirms the standard assumptions of diffusion theory about the importance of imitative behaviour for mass adoption. Erumban and Jong (2006) found that another indicator of value – 'power distance', discussed in chapter three – as well as uncertainty avoidance influenced the world-wide diffusion of PCs. This was found to be a robust result even after controlling for levels of education and income. Meanwhile, the analysis in chapter three of this volume found a real, though limited influence, of uncertainty avoidance and power distance on web 2.0 usage in Europe.

Communication cultures

Communication forms, patterns and expectations have been identified as potential influences in various writings on ICTs. One Finnish study by Puro (2002) raised the question of whether one can talk about a 'communications culture' consisting of expectations of appropriate speech behaviour (e.g. about the absences of small talk, the value of silence, the importance of being direct). That study discussed how these expectations were reflected in both fixed phone and mobile phone patterns of interaction, but also how the mobile might challenge traditional Finnish speech culture.

Other studies have distinguished between 'high context' or 'low context' communication cultures (Hall, 1983). In a low context culture most things have to be explicitly stated as people do not necessarily have a common understand-

---

9 Although one of the most cited works in social science citation indices, Hofstede's work is not without its critics – e.g. Fougère & Moulettes, 2006; Carbaugh, 2007.

ing of the context in which behaviour takes place. For Hall, France, and its proverbial 'on dit' ('one says'), was an example of a high context culture, whereas the United States, with its multiple ethnic groups, was an example of a low context culture. Such different contexts might help cast light upon patterns of communications in different societies.

In some countries it has been argued that it is the social control of certain forms of communication that shapes communication preferences. For example, Japanese researchers have argued that mobile e-mail in Japan was popular amongst youth partly because of the strong regulation of voice telephony in schools and public places (Okabe & Ito, 2005). Given 'no mobile phone' signs in trains and buses, and regular announcements over the loud speakers to this effect, almost no participants in this study made voice calls in these settings, but instead used mobile e-mail extensively.

Material culture

Finally we have material culture, where different cultural values have shaped and become embedded in the physical world, as reflected in the organisation of space (especially the rural-urban division), the styles of dwelling places, the types of item to be found in them and such matters as clothing fashions.

Arguably the national layout of urban centres reflects cultural influences as well as historical events. In single-node urban systems, the capital dominates the country such as in France and Britain. These encourage communication systems to be developed, deployed and used in different ways compared to multi-nodal system, such as in Germany or Switzerland. For example, the spatial concentration of potential mobile telephone customers in South East England facilitated the rapid roll-out of mobile telephony around the British capital while delaying that development in regions less attractive for the operators. In the multi-node case, every communication technology will tend to include a strong long-distance component.

The housing characteristics of different countries (e.g. size, interior design) vary. For example, in the 5-country qualitative study of telecommunications in 1996 there were differences between a number of European countries as regards the strategy of going to another room to seek privacy when making or receiving calls (Haddon, 1998). However, on further analysis this mainly reflected the distributions of different sized houses in the countries (and implicitly, different numbers of rooms). Comparing houses of the same size, many of the statistics differences disappeared, suggesting this search for privacy reflected the nature of the housing stock more than other values.

The spatial design of housing and the location of facilities within the home, is also a consideration. For example, in the UK (and other countries) the fact that in the early 20th Century many houses only had heating in the central room meant that people congregated there. Only later when other rooms were heated, and with the arrival of central heating, did children especially spend more time in separate bedrooms. This clearly might have some bearing upon where ICTs are used but also upon how they are used, given a lack of parental surveillance in these private spaces (as in current discussions of children's media rich 'bedroom cultures' in Bovill and Livingstone, 2001).

Lastly we have artefacts and cultural tastes. Here we might consider phenomena like fashion, cultural orientations towards creativity and novelty (in leisure activities, clothes and appearance) as well as the role of elite avantgarde culture. In other words, we can think of culture as expressed in areas of life such as art, decoration, design and commercial offerings. Such matters sharply distinguish Northern from Southern Europe, the latter being closer to Japan in terms of its fashion culture. This becomes all the more relevant when we consider that ICTs are not just functional artefacts but symbolic ones, which are as subject to the influences of fashion as are the other items we consume in everyday life. One need only think of the Nokia fashion mobile phones and the stylish PC colours now offered in some countries such as grey, black and silver computers in the US.

Conclusions

The intention in this chapter has been to be inclusive and fairly open-minded about ways in which we could see factors as being somehow 'cultural' in nature. No charting exercise would claim to be absolutely comprehensive, and indeed this is probably an impossible goal given that different researchers use slightly different definitions of what counts as cultural. Moreover, we acknowledge that boundaries around different aspects of culture are not fixed: the same examples conceptualised under one heading could, from a slightly different viewpoint, also fit under another.

Primarily, the chapter, and indeed the fuller report upon which it was based, can serve as a tool for making us sensitive to the wide range of ways in which factors conceptualised as cultural influence our experience of ICTs. This can help our attempts to understand differences in national patterns of diffusion, such as in analysis of the digital divide between countries. The chapter also reminds us that we have to ask about the specificity or generalisability of research conducted within a particular country. For example, if presenting country-specific research to an international audience, to what extent could those in other countries learn from it? To what extent might various factors, including cultural differences (but also the national histories of markets, the socio-demographic distribution of population, etc.) mean that findings are more or less likely to be replicated elsewhere? Finally, the chapter sensitises us to some problems and issues around defining culture, some of the limits of cultural analysis and areas where we might need to develop our thinking further.

Bibliography

Bovill, M. & Livingstone, S., 2001. Bedroom culture and the privatization of media use. In: S. Livingstone & M. Bovill. eds. 2001. *Children and their changing media environment. A European comparative study*. Mahwah, New Jersey, US: Lawrence Erlbaum Associates, pp. 179-200.

Calogirou, C. & Andren, N., 1997. Les usages sociaux du téléphone dans les familles d'origine immigrée. *Réseaux*, 82-83, pp. 187-203.

Carbaugh, D. 2007. From cognitive dichotomies to cultural discourses: Hofstede, Fougère and Moulettes in conversation, *Journal of Multilingual Discourse*, 2 (1). [Online] Available at: http://works.bepress.com/donal_ carbaugh/2 [Accessed 17 April 2010].

Carré, P., 1991. Un développement incertain: la diffusion du téléphone en France avant 1914. *Réseaux*, 49, pp.27-44.

Chung, L-Y. & Lim, S-S., 2005. From monochronic to mobilechronic. Temporality in the era of mobile communication. In: K. Nyíri. ed. *A sense of place. The global and the local in mobile communication*. Vienna, Austria: Passagen Verlag, pp.267-82.

COST 248 Mobile Group, 1997. Mobile telephony in Europe: histories, markets and modes of use. In : L. Haddon., ed. *Communications on the move: the experience of mobile telephony in the 1990s*. Farsta, Sweden: COST248 Report.

Craemer, P. & Franke, A., eds., 1935. Atlas des Weltfernsprechnetzes. Special edition of *Europäischer fernsprechdienst*. 5th ed.

Fischer, C. S., 1987. The revolution in rural telephony, 1900 – 1920. *Journal of Social History*, 21 (1), pp.5-26.

Fortunati, L., 1997. The ambiguous image of the mobile phone. In: L. Haddon. ed. *Communications on the move: the experience of mobile telephony in the 1990s*. Farsta, Sweden: COST248 Report.

Fougère, M. & Moulettes, A., 2007. The construction of the modern West and the backward rest: studying the discourse of Hofstede's Culture's Consequences. *Journal of Multicultural Discourses*, 2(1), pp.1-19

Gilligan, R. & Heinzmann, P., 2004. Exploring how cultural factors could potentially influence ICT use: an analysis of Swiss TV and radio use. In: L. Haddon, ed. *International collaborative research. Cross-cultural differences and cultures of research*. Brussels: COST, pp. 87-96.

Gilligan, R. Campbell, C. Dries, J. & Obermaier, N., 1998. *The current barriers for older people in accessing the information society,* First report of AOP-IS project. Düsseldorf: European Institute for the Media.

Haddon, L., 1998. Il controllo della comunicazione. Imposizione di limiti all'uso del telefono. In L. Fortunati., ed. *Telecomunicando in Europa.* Milan, Italy: Franco Angeli, pp.195-247.

Haddon, L., 2004. *Information and communication technologies in everyday life: a concise introduction and research guide.* Oxford: Berg.

Haddon, L. & Stald, G., 2009. Cultures of research and policy in Europe. In: S. Livingstone & L. Haddon, eds. *Kids online. Opportunities and risks for children.* Bristol: Policy Press, pp.55-70.

Hall, E. T., 1983. *The dance of life: the other dimension of time.* New York, US: Anchor Books.

Hallin, C. & Mancini, P., 2004. *Comparing media systems: three models of media and politics.* Cambridge: Cambridge University Press.

Headrick, D. R., 1981. *The tools of empire. Technology and European imperialism in the nineteenth century.* Oxford : Oxford University Press.

Heres, J. Mante, E, & Pires, D., 2004. Factors influencing the adoption of broadband mobile internet. In: Klamer, et al., eds. *ICT capabilities in action: what people do.* Brussels: COST, pp.133-154.

Hester, S. & Eglin, P., 1997. *Culture in action: studies in membership categorization, analysis.* New York: University Press of America.

Hoffman, D. & Novak, T., 1998. Information access: bridging the racial divide on the internet. *Science*, 280 (5362), pp.390–1.

Hofstede, G., 1980. *Culture's consequences: international differences in work-related values.* London, Sage.

Iske, S. Klein, A. & Kutscher, N., 2004. Nutzungsdifferenzen als indikator für soziale ungleichheit im internet, *Kommunikation@gesellschaft*, 5. [Online]. Available at: http://www.soz.uni-frankfurt.de/K.G/B3_2004_Iske_Klein_ Kutscher.pdf [Accessed 16 January 2010].

James, A. & Prout, A., eds. 1997. *Constructing and reconstructing childhood: contemporary issues in the sociological study of children.* London: Falmer Press.

Klamer, L. Haddon, L. & Ling, R. 2000. *The qualitative analysis of ICTs and mobility, time stress and social networking.* Heidelberg, Germany: EURESCOM P-903 report.

Lazarus, W. & Mora, F., 2000. *Online content for low-income and underserved Americans: the digital divide's new frontier.* Santa Monica, CA: The Children's Partnership, March. [Online] Available at: http://www. childrenspartnership.org/AM/Template.cfm?Section=Reports1&CONTEN TID=8194&TEMPLATE=/CM/ContentDisplay.cfm [Accessed 16 January 2010].

LeRay, E., 1994. *Communication et religion, la modernité et l'Islam dans un Québec vidéo-chrétien.* Mémoire de Maîtrise, Montréal, Canada: Université du Québec à Montréal.
Levine, R., 1997. *A geography of time.* New York: Basic Books.
Mante, E., 2002. The Netherlands and the US compared. In: J. Katz, & R. Aakhus, eds. 2002. *Perpetual contact: mobile communication, private talk, public performance.* Cambridge: Cambridge University Press, pp.110-25.
Miller, D. & Slater, D., 2000. *The internet. An ethnographic approach.* Oxford: Berg.
NTIA (National Telecommunications and Information Administration) & US Department of Commerce 2000. *Falling through the net: toward digital inclusion. A report on Americans' access to technology tools.* Washington, D.C.: US Department of Commerce and NTIA.
Okabe, D. & Ito, M., 2005. Keitai in public transportation. In: M. Ito, M. Matsuda & D. Okabe, eds. *Personal, portable, pedestrian: mobile phones in Japanese life.* Cambridge, Mass.: MIT Press, 205-18.
Pasquier, D., 2001. Media at home: domestic interactions and regulation. In: S. Livingstone & M. Bovill. eds. 2001 *Children and their changing media environment. A European comparative study.* Mahwah, New Jersey: Lawrence Erlbaum Associates, pp.161-78.
Portes, A., 1997. *Globalization from below. The rise of transnational communities.* WPTC-98-01. Princeton, US: Princeton University,.
Puro, J-K., 2002. Finland: a mobile culture. In: J. Katz & R. Aakhus. eds. *Perpetual contact: mobile communication, private talk, public performance.* Cambridge: Cambridge University Press, pp. 19-29.
Rainie, L. et al., 2003. *The ever-shifting internet population: a new look at internet access and the digital divide.* Washington D.C., US: The Pew Internet and American Life Project. [Online] Available at: http://www.pewinternet.org/Reports/2003/The-EverShifting-Internet-Population-A new-look-at-Internet-access-and-the-digital-divide.aspx_[Accessed 16 January 2010].
Rogers, E., 1995. *Diffusion of innovations.* 4th ed. New York: The Free Press.
Sautter, K., 1951. *Die Deutsche Reichspost 1871 – 1945.* Frankfurt/M.: Bundesministerium für das Post- und Fernmeldewesen.
SIBIS 2003. *SIBIS pocket book 2002/03.* Bonn: Empirica.
Simmel, G., 1976. *The metropolis and mental life. The sociology of Georg Simmel.* New York: Free Press.
Sundqvist, S. Frank, L. & Puumalainen, K., 2005. The effects of country characteristics, cultural similarity and adoption timing on the diffusion of wireless communications. *Journal of Business Research,* 58 (1), pp. 107-110.

Thomas, F., 1995. *Telefonieren in Deutschland. Organisatorische, technische und räumliche entwicklung eines großtechnischen systems*. Schriftenreihe Max-Planck-Institut für Gesellschaftsforschung, 21, Frankfurt/New York: Campus.

Thomas, F. & Mante-Meijer, E., 2001. Internet have and have nots in Europe. *e-Usages*. Paris 12-14 June 2001. ADERA: Bordeaux, pp. 411-426.

Thomas, F. Haddon, L. Gilligan, R. Heinzmann, P. & de Gournay, C., 2005. Cultural factors shaping the experience of ICTs: an exploratory review. In: L. Haddon, ed. *International collaborative research. Cross-cultural differences and cultures of research*. Brussels: COST, pp. 13-50.

Trompenaars, F., 1993 *Riding the waves of culture*. Antwerp/ Amsterdam: Economist Books.

Vehovar, V. Batagelj, Z. & Lozar, K., 1999. *Language as a barrier*. [Online] Available at: http://www.isoc.org/inet99/proceedings/3i/3i_3.htm [Accessed 16 January 2010]

Weiss, M., 2001. *Pisa 2000*. Berlin: Zusammenfassung Zentrale Befunde, Max Planck Institut fur Bildungforschung.

Yoon, K., 2002. Extending familialism through the mobile: young people's re-articulation of traditional sociality through mobile phones in Seoul, South Korea. *Third Wireless World Conference*. Surrey University, Guildford, 17-18 July, 2002. DWRC: Guildford.

Giovanna Mascheroni, Francesca Pasquali, Barbara Scifo, Anna Sfardini, Matteo Stefanelli, Nicoletta Vittadini

Chapter two. Young Italians' crossmedia cultures

Introduction

This chapter[10] discusses the findings of a qualitative study carried out by a workgroup of *OssCom (Research centre on Media and Communication)* and financed by MTV Italia. Its aim has been to investigate the development of new practices of media consumption and production among Italian young people (aged 14-24), in particular 'crossmedia diets', which use content and communicate across a combination of different media platforms and communication channels.

Such practices are clearly enabled by the diffusion of broadband, mobile media, digital television and the media convergence processes between institutions and at production level. Nevertheless, these practices are also influenced by social and cultural factors such as age, gender, household composition, the size of one's social networks, and so on.

The study has followed a multi-sited approach, utilising several investigative techniques (in-depth interviews, participant observation with the support of visual sociology and an exploration of Italian online discussion sites).

The findings related to teenagers' and young-adults' consumption practices presented here draw on a perspective that highlights some specificities of the current generation of Italian youth in terms of their development of cross-platform consumption diets, and in relation to several variables (such as the spatial-temporal contexts of consumption, the role of the peer networks and the media products). In particular, we will focus on screen-based media consumption and technologically-mediated interpersonal communication.

The findings about Italian youth's media cultures provide a chance to reflect upon some relevant issues within the contemporary debate about media

---

10 This chapter is the outcome of the discussion involving Anna Sfardini (who mainly wrote the first two sections 'Media in transition' and 'Overview: media change in Italy'), Matteo Stefanelli (the sections 'Television facing convergence' and 'Screen-based media consumption'), Francesca Pasquali (the sections 'The research project' and 'Conclusions'), Giovanna Mascheroni (the section 'Methodology'), Barbara Scifo (the sections 'Media competition in crossmedia cultures' and 'Technologically-mediated interpersonal communication') and Nicoletta Vittadini (the section 'Media integration in crossmedia cultures'). The authors thank Massimo Scaglioni, Antonio Dini and all the junior team of researchers for their contribution at the research.

convergence: especially, the relation between private and public contexts of consumption, between mobile and domestic media, between social broadcasting media and networking social media, and between linear and non-linear ways of consuming media.

Media in transition

We are currently witnessing a phase of change, characterised by considerable scope for negotiation and by unpredictability. Change in media offerings (concerning the redefinition of the role of market players, various forms of economic valuation of media products, and even issues of standards and copyright) goes together with increased possibilities for, and new forms of, media consumption.

The ubiquitous nature of digital and networked media (from the multimedia internet to personal media and ICTs) defines an ever more articulated and complex scene wherein subjects move and make their choices (Ito, 2007). In this scene analogue platforms (e.g. general television) sit alongside various new digital platforms (e.g. digital, multi-channel and multi-thematic television). Traditional forms of distributing television, films or music (e.g. broadcasting) exist alongside new circulation practices (e.g. by p2p online networks). One-to-one modes of communication (e.g. fixed or mobile phones) sit alongside many-to-many communication channels (e.g. Instant Messaging, blogs, social networking sites). Content produced by institutional and commercial actors exists alongside user-generated content of an ever more multi-medial nature (containing not only text, but also pictures and videos). Niche content, to be shared within a closed social circle, is to be found alongside mainstream content. And so on.

The increasing complexity of this contemporary mediascape makes it absolutely urgent to reconceptualise media change more broadly, beyond the usual utopias and dystopias. For some years now the need to address the dynamics of media convergence beyond its mere technical dimension has gained prominence across the theoretical literature (Gitelman & Pingree, 2003; Jenkins & Thornburn, 2003a; Jenkins & Thornburn, 2003b; Jenkins, 2006; Gitelman, 2006). The digital revolution paradigm, oriented more towards the transformative features of digital technology, is thus being replaced by the *convergence paradigm*. This paradigm appears more responsive to the multi-dimensional nature of media change, to the role of people alongside that of technologies, and to the mechanisms of hybridisation and re-mediation operating between old and new media. In Jenkins' terms:

"convergence represents a paradigm shift – a move from medium-specific content towards content that flows across multiple media channels, towards the increased interdependence of communications systems, towards multiple ways of accessing media content, and towards ever more complex relations between top-down corporate media and bottom-up participatory culture." (Jenkins, 2006, p.243).

This attention to the forces and the actors shaping media technologies is one of the striking features of this convergence paradigm. While the advent of new media and the process of digitalisation provide the conditions for widespread change within the media system, these same conditions are being actively shaped by the various actors populating the contemporary media environment. These are the multimedia conglomerates (on the supply side), public institutions (on the governance side) and users themselves (on the consumption side). A medium, therefore, cannot be defined unless one starts from its accompanying 'protocols' and 'practices', which shape it on the cultural, economic and social level (Scaglioni & Sfardini, 2007).

The present global phase of media change is characterised by tactical decisions and unforeseen consequences, multiple signals and competing interests, and above all by uncertain directions and unforeseeable outcomes. The dynamics of convergence assume different characteristics depending on the specific cultural and economic contexts, which in turn depend upon the various national histories of media systems. We will attempt to provide a (non-exhaustive) depiction of the Italian context, starting from a snapshot of the diffusion of different digital platforms.

Overview: media change in Italy

While analogue television appears today, together with radio, to be the platform with the highest penetration rate among the Italian population, an examination of the technological equipment actually owned by families reveals highly differentiated diffusion rates among the four typologies of devices and services (ICTs, informatics, entertainment and consumer electronics) upon which digital convergence is centred. As of June 2006, according to data from the Ministry of Communications, the mobile phone remains the most diffused technological device, being present in more than 80% of households (two or more per family, in the majority of cases).

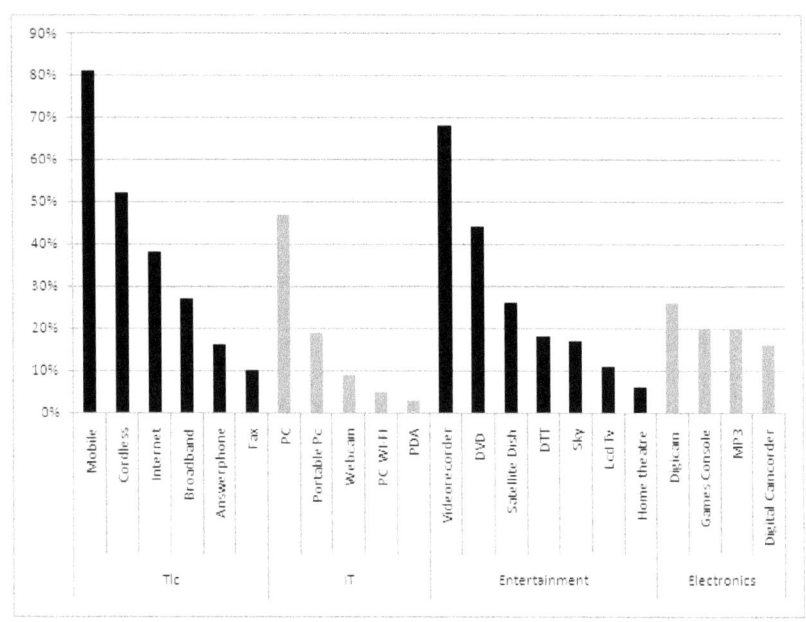

Figure 1: Technology diffusion among Italian Households.
Source: Ministero delle Comunicazioni, Osservatorio Banda Larga (2006).

The personal computer's diffusion is limited to less than half of all families (47%), albeit growing at more than 3% a year since June 2004. However, with regard to the PCs, market trends have led to two substantial transformations: in 2006, laptop sales (2,540,000 units) exceeded those of desktop PCs (2,255,000), and the total number of PCs in Italy (counting both workplaces and households) reached 24.7 millions (data from Assinform 2007 Report). Apart from the diffusion of the above mentioned platforms, in the last few years technological innovation has been driven by new ICT products as well as by the development of entertainment based consumer electronics. As regards ICTs, the highest growth rates have been recorded by UMTS[11] and broadband-enabled mobile phones, which finally managed to overcome their niche status (by the end of 2005 UMTS users numbered almost 10 million).

---

11 Universal Mobile Telecommunications System is one of the third-generation (3G) mobile telecommunications technologies.

Broadband diffusion rates look stunning. By September 2006, on the infrastructural side, xDSL[12] (the current technological reference point for broadband) was available to 88% of the population. On the diffusion side, records show a growth of 20% since June 2004 – broadband is to be found in 27% of Italian households (and in 38% of internet connected households)[13]. With regard to consumer electronics, the fastest-growing devices were DVD video, DTT (Digital Terrestrial Television) recorders, digital cameras, digital camcorders, and, more recently, MP3 players.

Television facing convergence in Italy

Within this context, the most central medium – television – is itself undergoing transformations that are redefining both its role in the wider media scene and the positioning of its players with regards to the development of the various digital platforms (Aroldi, et al., 2006; Colombo & Vittadini, 2006). In particular, forecasts predict a strong diffusion of DTT, matching the progressive extinction of analogue television that ends in 2012. This is the date set by the Italian Government for the switch-off, which is going to involve almost twenty-four million national households. However, these forecasts look optimistic, considering the shortcomings still affecting terrestrial digital television.

The satellite platform, which showed the highest growth rates during the period 2003-2007 thanks to the launch of SkyItalia, should stop growing due to the saturation of the pay-TV market, stabilising at around five million households.

IPTV (TV over the internet) diffusion is related to the growth of broadband connections (estimates predict more than 15 millions in 2012), and although it only began in 2007 it should become a market ripe with offerings and contents. Here too estimated growth rates – supposed to surpass satellite subscriptions around 2011 – may appear to be too optimistic. Broadband diffusion might naturally relate to, and drive, the diffusion of various forms of internet TV, outside the "closed gardens" of IPTV. Claims about the future diffusion of TV on the mobile phone should be approached with even more caution, since it is affected by further uncertainties concerning the prevalence or the co-existence of a linear flow-based TV (based on the DVB-H[14] platform) or of audiovisual on-demand services (based on UMTS and Hsdpa[15]).

---

12 xDSL Refers collectively to all types of digital subscriber lines, a family of technologies that provides digital data transmission over the wires of a local telephone network.
13 Data from Osservatorio Banda Larga. Available at: www.osservatoriobandalarga.it [Accessed 13 January 2010].
14 Digital Video Broadcasting - Handheld is a mobile TV formats.
15 High-Speed Downlink Packet Access is a protocol for mobile telephone data transmission.

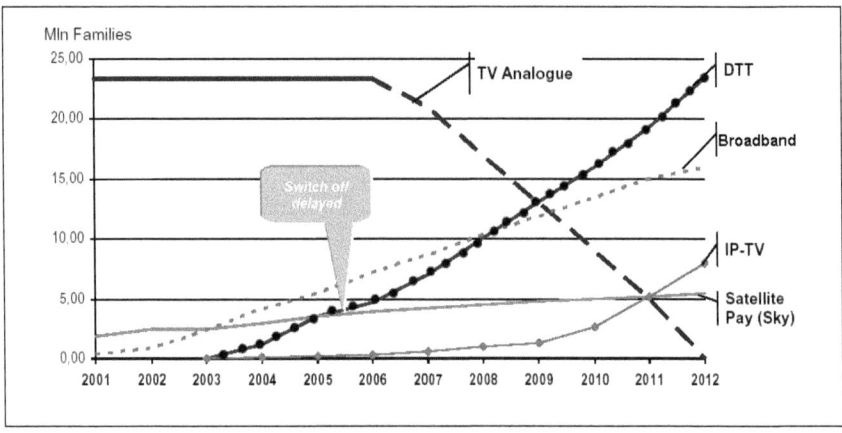

Figure 2: The diffusion of digital platforms for TV contents among Italian households (Mln). Source: Mediaset/Booz Allen Analysis, based on forecasts by Telecom Italia, Informa Media, Gartner, ScreenDigest, Sky and Ministero delle Comunicazioni

The research project

Despite some delays, Italy has finally entered a fully networked information economy, characterised, in the terms of Yochai Benkler by the presence of a

> "communications environment built on cheap processors with high computation capabilities, interconnected in a pervasive network." (Yochai Benkler 2006, p.3)

Corresponding to this change at the technological and institutional level we are starting to find a re-definition of media consumption, at least in those audience niches that are more easily reachable by the processes of technological innovation. In the Italian case, the most recent data on media consumption indicate that adolescents and youths are the population segments more directly involved in the emergence of the networked information economy. A key example of this is the growth among Italian adolescents and youth of downloading and file-sharing practices (Bennato, 2007) and of web 2.0-related services[16] involving a convergence between institutionalised and user-generated content as well as an inter-connection between different communication platforms.

---

16 Data published by Nielsen//Netratings, in April 2007, state that in the previous few months 56% of Italian web surfers used web 2.0 sites, with a particularly strong diffusion among younger users. As a result, Italy ranks fifth among European nations in terms of the traffic on sites such as YouTube, MySpace, and Wikipedia (Nielsen//Netratings 2007).

Italian youth and adolescents are therefore now fully integrated within this *convergence culture*

> "where old and new media intersect, where grassroot and corporate media collide, where the power of media producer and media consumer interact." (Jenkins, 2006, p.2).

These structure new media diets.

While the situation benefits from the relevant technological variables we discussed in the previous section, these must nonetheless interact with the specific social and personal variables that, according to social shaping theories of media and ICTs[17], drive the adoption and incorporation of technologies into everyday life.

The research whose results are presented here, was designed to study the emergence of new cross-media diets among young Italians. While this group remains a consumption niche, it nevertheless appears to provide a particularly interesting case study. This group domesticates digital technologies rapidly. Young Italians participate in large social networks that act as reference groups wherein media contents – even traditional ones – are important resources for socialisation (more so than for other age groups). Finally, they have a considerable amount of time available for both media consumption and technologically-mediated or face-to-face relationships.

Methodology

The complexity of the chosen topic of research led to a rethink of traditional ethnographic approaches to studying media consumption and required the adoption of a flexible and multi-sited methodology (Marcus, 1995). In a way analogous to the line of thought that originated in the field of internet studies – and especially in internet ethnographies – the methods of 'mobile' ethnography were adopted (Hine, 2000; Hartmann, 2006). This methodology captures the inter-connections between life (both on- and off-line) and the flows of interaction, thus requiring researchers to conduct their studies in multiple sites and with regard to multiple research objects, in this case multiple media and technologies.

Being both adaptive and reflexive and able to progressively redefine itself on the base of the research findings as they emerged, this approach allowed us to set up an iterative interaction between the empirical data and the theoretical basis of the research, and to triangulate several tools of inquiry and objects of

---

17 For further discussion of these topics, see the section on *Technology Design and Development* in Lievrouw & Livingstone (2002) and Haddon, et al. (2005).

study. The research was articulated across three areas that were in a continuous, reciprocal dialogue. Each area corresponded to different objectives and engaged different inquiry techniques: in-depth interviews and both on- and off-line participant observation.

The triangulation between these different research tools and observational environments was intended to reconstruct the relationship between Italian youth and the various platforms they use, insofar as they are integrated into the domestic and extra-domestic contexts in which they operate and are shaped by the daily lives and individual biographies of these youth.

The processes through which new digital media platforms were incorporated in young people's lives within extra-domestic contexts were revealed through ethnographic observation at the places where urban youth meet and are entertained (a gym, a shopping mall and a games arcade). This was captured through ethnographic notes, photos and video. Here the nomadic consumption practices of these youth and their contexts in which they use media and socialise could be observed. Ethnographic observation was integrated with brief, non-directive individual and group interviews with the young people frequenting these places, in order to reconstruct the discourses and dynamics of outdoor digital consumption.

In addition, 40 non-directive in-depth interviews were conducted to explore their domestic media consumption. The interviews were conducted in young people's homes, in order to observe their social context, and the devices they owned, the physical collocation of these devices, in general to explore better the 'moral economy' of the household regulating access to technologies (Silverstone and Hirsch, 1992).

The choice of the sample interviewed, divided equally between males and females, was based upon the combination of two main variables: the typology of the platforms owned and used by the young people and the age of those youth. With regard to the first aspect, the sample was segmented on the basis of the ownership of devices from at least two of the following platform macro-typologies: digital television platforms (DTT decoder or satellite TV or IPTV); PC with a broadband connection; mobile devices supporting audiovisual practices (third-generation mobile phone or TVPhones or portable players, such as MP4 players and Sony PSPs). With regard to the age variable, the sample was distributed as follows:
- 5 pre-adolescents (11-13 years of age) and 9 adolescents (aged 14-18), meaning a total of 14 people.
- 15 youths[18] (19-24 years of age) and 11 young adults (aged 25-35, among which 6 were under-30s and 5 were over-30s), making a total of 26 people.

---

18 Different countries categorise 'youth' differently. In some countries the age group 18-24 year old would be considered to be 'young adults', but in Italy they are seen as being 'youth'.

In this chapter we will focus on the empirical data gathered through home interviews and in particular on the media consumption practices of adolescents (aged 14-18) and youths (aged 19-24). The aim is to highlight the emergent forms of consumption across the different platforms in the early stage of their incorporation into daily life (a phase characterised by high uncertainty and unpredictability).

Media competition in cross-media cultures

Within the above-depicted multi-platform media environment, young users are faced, on the one hand, by ever more interconnected and inter-operating multimedia and multitasking digital technologies. On the other hand, they face a set of technologies and services allowing for the same media consumption practices (e.g. watching films, listening to music, or communicating through mediated interpersonal channels) but with different modalities, timings, contexts of use and formats. Our focus, then, shall be on understanding not only how these practices are enacted within the new media context, but also on the reasons driving the choices made between the many options (both old and new) that are possible today, as well as on the specific and differentiating features of each practice. For a similar approach applied to the practices of mediated interpersonal communication see Haddon (2005); for one focusing on screen-based media see Livingstone (2002).

What factors, not only technological in nature but also economic and above all socio-cultural, define the selection criteria and make the various ICTs at times compete with each other as alternatives and at other times complement each other? How do these selection processes impact upon the way digital and networked media are socially incorporated into life and upon the way that preexisting media and communication practices are symbolically and functionally re-configured since, as Jenkins points out:

> "old media are not being displaced. Rather, their functions and status are shifted by the introduction of new technologies" (Jenkins, 2006, p.14).

The assumption upon which this article is based, part of a long tradition within media studies, is the unacceptability of the "displacement" thesis, according to which the arrival of a new medium implies the displacement of pre-existing ones. Rather, the experience of pre-existing media seems to be itself modified when accompanied by the new medium, thus generating a more differentiated and specialised pattern of use, increasing the complexity of the media mix. For an overview of the debate concerning the relationship between old and new media, see, among others, Pasquali (2003).

These questions become crucial when the ubiquitous nature of digital and networked media (from the multimedia internet to personal media and ICTs) define an ever more articulated and complex landscape in which people exercise their choices.

The purpose of this paragraph is to re-read the research data findings in order to highlight the processes of selection that youth make at three different levels, corresponding to the moments of *appropriation* of the available media and contents (the choice of whether to buy or not to buy a ICT), of *access* (the choice of which ICT to use from among those that are owned) and of *use* (the choice of how to use the chosen ICT, i.e. which practice). In particular, we shall focus our attention on some socio-cultural factors shaping these choices, as documented in chapter one, such as the spatial-temporal structures of daily life, social networks, individual biographies and gender. However, within this wide range of recorded practices, we shall here focus only on screen-based and communication-related practices.

Screen-based media consumption

With regard to the practices of screen-based media consumption, we can first distinguish the platforms from their contents. As regards the platform, the centrality of television is declining for this age segment as linear television consumption practices (e.g. watching live broadcast TV programmes) compete with non-linear audiovisual consumption practices (e.g. accessing the material on demand). The old general, analogue television broadcasting still retains some value, but it must compete with digital television platforms, other media platforms and communication platforms, as regards the above-mentioned dimensions of appropriation, access and how the material is used or viewed.

As regards the process of *appropriation*, the competition between television and other media is first and foremost driven by domestic negotiation with other family members, still highly influential for this age segment, and also by economic factors. The adoption of digital platforms such as the personal computer, for example, occurs within a set of intra-family dynamics whereby the purchase of a PC often takes place for a broad spectrum of reasons centred around its collective utility for all the family and its value as an educational tool. Meanwhile, the greatest investment made by young people as individuals, in terms of their personal adoption, is in mobile devices. The limited amount of money available for spending and the lower relative cost of such technologies together help shape the decision to buy basic MP3 players and mobile phones, without any interest in the more advanced standards (DVB-H mobiles are notably absent from this age group). This process of selecting the platform is also influenced by young people's peer networks, which strengthen the social relevance of those ICTs that are considered to be more strategic in their lives (i.e. mobile phones, MP3 players and tools for online communication).

A typical example showing the importance of domestic negotiation can be seen in the adoption of additional television platforms – especially, satellite TV. The presence of SkyItaly in the household often originates from intra-familial power dynamics, and its acquisition depends upon the politics of purchasing. In this process, youth usually assume a secondary role, or else they are have a weaker negotiating position compared to the influence of both parental consumption choices and whether children's or pre-adolescent contents are widely available on the Sky channels.

As regards *access* (i.e. the decision about which available ICT to use) the competition between television and other media is very harsh and is influenced by several factors: among them, time, space and social relations. Foremost is the general decline in the amount of free time spent on domestic activities, which leads to television consumption being relegated to a residual position in life (especially among young college students and/or workers). The temporal economy of daily life influences access to screen-based platforms, for example decentring those platforms with the lowest symbolic investment (such as analogue TV). It also leads to a re-valuation of other platforms within specific contexts (such as viewing music videos on mobile phones or iPods when travelling, when studying and when on vacation).

The spatial dimension is another factor that may influence the logics of access, driving platform scarcity or abundance in specific places or domestic spaces. For example, the presence of satellite TV in the living room can have a bearing upon the selection of analogue TV as the secondary platform, and it may even have a bearing upon on the selection of films to be consumed on a PC (in young people's bedrooms) as an individual practice involving viewing contents that have not been negotiated with parents.

An individual's relationship with other people in his or her social networks also appears highly relevant in shaping his or her use of various media. On the one hand, the desire to communicate with these peers pushes young people towards favouring CMC (Computer Mediated Communication, such as Instant Messaging) and mobile telephony over cinema and television. On the other hand, the interest in interacting with others in those networks can lead to more value being conferred on those screen-based platforms that offer the video contents with the highest 'social spendability', (i.e. those that might be most appreciated by peers), such as audiovisual files that can be converted into items that can then be traded online with other network members. This happens both with those audiovisual files containing narrative products (such as films and TV-series), which file-sharing practices thrive upon, and with short video files ('amusing videos', user-generated-contents), which provide material to be exchanged via email, chat and Instant Messenger. Familiar gate-keeping roles can be found here. Within this younger population such gate-keepers influence access to certain platforms in terms of timing, costs and contents. Meanwhile, the more competent youth (usually males) or older peers (often older brothers) are designated as being leaders as regards the downloading of music and audio-visual files.

As regards the dimension of *use*, in the case of audiovisual consumption the platforms that are usually used for ritual family practices are in direct competition with those that are used for individual practices. While analogue TV and even satellite TV viewing (or DVD viewing through a DVD-player) are integrated into the dynamics of rituals internal to the family (or to friendship relationships), watching DivXs[19], or films or TV-series that have been downloaded, is an individual practice. Furthermore, it should be noted that the very marginalisation of television consumption can itself be traced back to the intensive use of the personal computer and of the internet as platforms allowing access to those very same television contents. In this sense, downloading a TV-series or films, and to an extent streaming funny videos and UGC[20]s, are for these young people practices that are competing with the traditional television or DVD use.

Meanwhile, the linear viewing of television, either via the old analogue platform or via the new digital ones, is for some people still a central practice that provides a reference point for then choosing contents that can be consumed in a non-linear fashion (as when locating TV-series, films or music videos that can then be retrieved and consumed on other platforms).

This dimension is also influenced by factors tied to the subjects' social networks, which uphold and drive the discursive centrality of some television contents (i.e. discussions with peers about brands, themes or characters). On the supply side, the general competition between analogue and digital TV (mostly satellite and IPTV, but also to some extent the DTT platform) revolves precisely around the contents being offered. While there is a tendency for youth to symbolically de-invest themselves from analogue TV, digital satellite television is one platform that is re-activating audiovisual consumption. Access to audiovisual content is further influenced by factors relative to young people's time budgets. For example, interstitial times – breaks in study, moments of socialising at school, journeys to school or work – influence how they consume short videos (funny videos, UGCs). Meanwhile, free time in the evening leads to the consumption of specifically-narrative audiovisual contents (films, TV series).

---

19 DivX is format that allows video compression.
20 User Generated Content.

Technologically-mediated interpersonal communication

Within their media diets, Italian adolescents and youth make strong investments of both time and identity in technologically-mediated interpersonal communication practices (either via mobile phone or via the internet). These practices often colonise time previously dedicated to other media-related (or non media-related) activities (such as watching TV or simply filling up daily interstitial times). Furthermore, the management of technologically mediated interpersonal communication is a terrain of fierce competition (essentially between voice calls, SMS, Instant Messaging, e-mail and VoIP), which has gone through a major reconfiguration. This reconfiguration started with the availability of flat-fee, always-on broadband connections, which actually re-defined the role previously played mainly by mobile phones (Scifo, 2005). Thus mobile phones effectively compete with forms of computer mediated communications, which, in turn, arise from an interaction between the nature of the spatial-temporal configuration of the communication itself, the other communicators and the communication's content and purpose.

Let us start, then, with the use of the broadband-connected PC with Instant Messaging software (mainly the highly popular Microsoft Messenger, MSN). Its diffusion and ubiquity among those young people who can access this platform represents the most recent development on the contemporary Italian scene as regards interpersonal mediated communications.

The *appropriation* of instant-messaging software is clearly made possible by the availability of always-on broadband, whose fee is paid for by the family and not by the young user. But the decision by young people to adopt this platform is socially driven by the pressures of the peer network. These pressures play a fundamental role in the diffusion of knowledge about how to use this tool, they fuel the drive towards the adoption of the software as a symbol of group belonging, and, finally, they give rise to and support the relevant literacy skills and social learning processes.

However, while young people may be in possession of the software, *access* cannot not be taken for granted. Its times and spaces are negotiated with other family members, particularly at the intra-generational level (i.e. among siblings). This is due to the limited number of internet-connected PCs in households (typically featuring a single unit in the children's – often shared – bedroom).

Whenever accessibility is granted, the actual *use* of the MSN software is shaped by several factors. First of all, from a temporal point of view, a central factor is that young people need to synchronise with the times when peer networks are online, times which are in turn determined by the alternation between two different spatial-temporal contexts. The first relates to the relation between social obligations (at school or university) and outdoor pleasure time. The second involves being present in the home during times when the PC becomes accessible (e.g. in the morning before leaving home, during the afternoon and/or the evening when the Instant Messaging can be used concurrently either with other study-related activities or with PC-related ones such as web surfing, downloading and listening to music, and watching television).

With regard to the interlocutors, Instant Messenger is directed at the young person's close and bounded social circles. That is to say, it sustains pre-existing networks of relationships instituted through daily life[21], but limited to the peer group (to classmates, to out-of-school friendship networks), thus excluding adults (typically parents). Therefore, this social norm, shared among adolescents and youth, deems instant messaging to be a practice typical of these age cohorts (indeed, findings show reduced adoption and usage of Instant Messaging in pre-adolescents and a greater predilection for email among young adults).

Finally, as regards the contents and purposes of communication, the use of MSN operates at several levels: the phatic one (i.e. simply maintaining some contact), its use for social micro-coordination (e.g. organising the logistics of when to meet), for sharing and for allowing young people to express their identities. It is a truly multi-functional tool whose uses go beyond mere synchronous communication (that is, chatting). As a background practice (either alongside other computer-related practices or as a form of dedicated communication), it is instrumental in the organisation and synchronisation of the activities of outdoor informal sociality. Furthermore, it plays an important role in terms of supporting the exchange of multimedia materials (both personal and otherwise), such as photographs, music files and links to popular funny videos, as well as through building spaces for self-expression and sharing them with the restricted peer network, through the blog area offered by MSN.

Two developments need to be noted. The first involves the consequences of the rise of the Instant Messaging practices for the re-configuration of young Italians' pre-existing mediated (and non-mediated) communication practices. The second relates to the kind of competitive dynamics it generates between the various possible options.

First of all, young people's considerable temporal and symbolic investment in Instant Messaging makes it one of the main drivers of PC use (indeed, the use of MSN is a very first action that takes place right after powering up of the unit). But above all, IM is an agent for the symbolic re-definition of the PC, re-domesticated through IM so that it becomes a networked 'ego-centred' platform (and not only as a platform for gaming, study and multimedia use).

The second development, even more relevant, is the way in which the use of Instant Messaging re-configures that of SMS, its direct competitor. Indeed, MSN and SMS exchanges take place within the same social networks, but since in households access to both platforms is available, MSN's competitive edge

---

21 From this standpoint, MSN has completely displaced open chatrooms.

results from its economic advantage (no charge for young people, who exploit a family resource and not a personal one), from its technological features and capabilities (writing via the PC keyboard is more efficient and comfortable) and because it supports more easily many-to-many multimedia communications.

In which case, why, when and in what forms does SMS usage continue to exist? First of all, within the domestic context the use of SMS acts as a substitute when online connectivity is simply not a possibility (e.g. if the interlocutor does not possess broadband technology, the software or the rights to access it). Furthermore, the decision to use SMS depends upon the kind of communicative content to be conveyed and upon the different social location of users. SMS still plays a fundamental role in the ultra-short-time micro-coordination of users (be it 'just in time' communications or outdoor ones).

Finally, the incorporation of MSN in young people's lives has re-defined the use of email (more frequent in young adults than in adolescents). Email is deemed to be a slower means of communication, whereas the communicative needs of youth hold speed to be fundamental, including for the above mentioned short term micro-coordination. Hence, always-on connections and a familiarity with MSN lead to the constant monitoring of the electronic mailbox when at home. However, email has a competitive edge when there is a need for communication that does not have to synchronised, in order to manage the planning of time either for co-ordinating study or work activities or for social coordination over the long term. Furthermore, from a symbolic point of view, email is perceived as involving fewer social obligations and being less invasive.

In closing this section, we shall only briefly mention the other digitally mediated communication practices that can managed through the computer or through mobile devices. As regards the PC, young Italians greatly appreciate video-calls made possible by VoIP systems (e.g. Skype, which involve voice communications via the internet). Immediately perceived as a video communication service, these VoIP systems are starting to compete with MSN. In contrast, there is significantly lower interest in mobile video communication. This is often viewed negatively (due to the lack of privacy during the act of communication), especially when it is compared to domestic video communication (through a webcam connected to either MSN or to VoIP). Finally, no intensive use has been recorded with regard to MMS[22]s (sending images directly form one mobile phone to another). Whereas the simple act of taking photographs via the camera phone appears to be highly appreciated and widely adopted, this is generally followed by an exchange of materials via MSN or e-mail, rather than via MMS.

Therefore, while Instant Messaging is becoming the main support platform for digital media communication, SMS and email still co-exist as the main

---

22 Multimedia Messaging Service.

alternative tools, whose use is differentiated in relation to the interlocutors' contexts and to the communications' contents.

Media integration in cross-media cultures

Cross-media diets stretching across different technological platforms and contents are integrated into the daily lives of Italian youth and adolescents through the practices of acquisition, consumption and the exchange of those contents. Certain cultural factors can act either as drivers of or as constraints on these practices of crossmediality. These factors include the temporal structures of daily life (societal structures shaping time use as well as cultural expectations about that same time use), the technological configuration of the households and values (both of the family and of social networks).

Meanwhile, the presence of multi-functional media technologies (PCs, mobile phones and other mobile devices) enables 'multitasking' consumption practices using the same platform (e.g. using Microsoft Messenger while listening to music files stored on the PC and/or sharing audiovisual files; or playing games while listening to music on mobile phones). Although this multitasking appears to be dominated by a *'centripetal logic*' (the *logic of convergence*), as regards what has been called 'networked public' (Ito, 2007) practices in Italy respond to different logics, opening up the possibility of an evolution towards new scenarios. We will describe in this section the cross-platform mobility of contents (i.e. how contents move across different platforms), the cross-platform acquisition and the consumption of content and the cross-platform cultural exchange of either institutionally-produced or user-generated contents.

The technological configuration of households, involving the presence of multiple platforms (e.g. analogue/satellite TV, personal computer), and the strong affection (close to fandom) for certain television brands and content (e.g. The Disney Channel, MTV; serials, cartoons and music videos) both act as drivers developing cross-media mobility in two directions:

- We have the prolonging or extension of the consumption experience, by repeating it either through switching consumption practices between digital and analogue television platforms or through accessing stored material (see, for example, the Sky reruns of TV-series or the archiving of cult TV-series on DVD or DivX). This is coupled with the intensification of the experience through gathering related materials on the internet (either from institutionalised sources or from other users' 'grassroots' sources).
- Then we have a kind of substitute consumption (even in relation to contents with low symbolic investment) achieved through the retrieval of the synopses of missed episodes or through the podcasting of segments of the beloved

programmes, in cases when it is not possible to access the appropriate analogue or digital television platforms.

These forms of consumption grow in relevance with the age of users (being more present in youth than in adolescents). This is due to older youth having a different orientation to web 2.0 (and the accompanying literacy) as well as to the fact that they ascribe a higher value to breaking free from the rigidity of linear, flow-based schedules. In these cases, cross-platform mobility acts according to a '*centrifugal logic*', whereby a central medium (TV or cinema) remains dominant but nevertheless prompts additional uses of other media and technological platforms as 'accessory uses' aimed at enhancing the consumption experience itself.

Even more relevant among the young Italian 'networked public' is the development of cross-media content acquisition and consumption. Indeed, the cross-platform diets of young Italians generally feature patterns structured into three phases:
– Locating and consuming contents (e.g. music files, films, TV-series) that can be found on mainstream media platforms (on analogue TV, radio, in the cinema) or suggesting contents to each other through word of mouth (either virtual or in real life).
– Acquiring content either through internet file sharing and then streaming it (with initial, often transitory, storage on hard disc) or through the purchase of originals (e.g. music CDs and DVD boxed sets).
– Transferring that content onto other forms of storage (DVD/DivX, CD) or onto mobile forms of storage (MP3/MP4 players, Sony PSPs[23], mobile phones) either for consuming at other times and in other places, for archiving, or for later trading or exchanging.

The relevance and intensiveness of these practices among young Italians is supported by several factors related to the evolution of the spatial-temporal structures of consumption and involving their orientation to the family and the wider group.

The first is the value that young Italians ascribe to having some flexibility in their ability to consume due to the increasing limitations of the spatial-temporal structures in which they operate. Indoors this related to the ever-demanding negotiation for access to digital platforms (e.g. access to television and the PC). Outdoors this is related to the progressive emergence of new spaces and times for the consumption of media products both individually (during increasing interstitial or 'in-between' times during routine patterns of mobility) and collectively (with the progressive expansion of the time spent meeting social obligations – e.g. school – combined with time set aside for informal sociability, such as time allotted for the peer group).

---

23 Play Station Portable – a games console.

The second is the large amount of disposable time available to this age group (especially in the case of students). This includes a large amount of time spent indoors (typically study time) into which file sharing fits easily as a background practice while accomplishing other activities on the PC, or while consuming others' audiovisual contents (when they are not colonising the night). In fact, the decrease in free time that occurs when people reach working age leads to a progressive reduction in the frequency of file sharing, more selective file sharing, and a change in the perception of file sharing, now seen more as an excessively time-consuming activity.

The third is the relevance of friendship networks. These both motivate young people to use services and software as tools for symbolically marking their group participation and their identity and these social networks are also contexts for socially learning how to use file sharing software and how to develop related practices.

However, cross-platform content acquisition and consumption are influenced by a combination of social networks and spatio-temporal constraints. The scarcity of household PCs within the home makes usage times dependent on the platform's accessibility to family members. Meanwhile, unequal degrees of literacy and levels of skill among youth as well as within the micro-context of the family can also lead to various processes:
– younger people (adolescents) experience gate-keeping limitations on the consumption whereby either older (typically male) siblings exercise forms of control over software access, or parents monitor the timing and cost of connections.
– older youth delegate the management of access to more competent local experts (brothers, boyfriends, friends – also typically male).

In these cases, cross-platform practices act according to what may be called a *'linear logic'*. The orientation provided, or agenda setting role played, by TV and/or radio platforms appears to be highly relevant as a driver of these cross-platform practices, especially in relation to content selection. This is the logic pursued by 'young' brands and channels, and by some formats (such as films, TV-series, or music products – even non-audiovisual). These products are consumed independently from the broadcasters scheduling, or else configure themselves as cult products and are therefore purchased (a) as collectable items (b) for recovering contents that evoke emotion or re-enact memories, (c) for pursuing non-mainstream materials or (d) for screening materials in advance.

Finally, another development to note is the cross-platform cultural exchange of both institutionally-produced and user-generated contents (UGCs). Among youth, media contents live their particular social life through a *'network logic'*- driven by circulation across various groups. Exchange patterns are sustained by networks of relationships instituted through mediated communication (e.g. Microsoft Messenger, Fastweb). Here contents themselves may circulate (e.g. as music files, trailers, films, and UGCs such as pictures and videos) or the web links may be shared. Or they may be traded as items within real-life networks of relationships, through lending or being given as gifts stored digitally (on DVD, DivX, CD) in the case of films and TV series, or through wireless data exchanged between mobile devices (such as MP3/MP4 players, mobile phones, Sony PSPs) in the case of music files, amusing videos and UGCs. Different types of content are also uploaded, although this practice is less common, onto personal web spaces (e.g. blogs, social networking sites).

In all of these cases, content exchange is central to these highly trans-platform practices (e.g. UGCs produced through mobile phones and stored on PCs, or institutionally-produced contents acquired through file-sharing and stored on DVDs and DivXs). Within such patterns of exchange, contents acquire a value in terms of their contribution to maintaining social capital, as noted earlier they have a social spendibility, which gives rise to high levels of identity investment (e.g. in UGCs, music files). In fact, even low-investment, spontaneously consumed contents such as amusing videos (acquired from YouTube and Google Video) are nevertheless valued as tools for the preservation of both real-life and virtual social networks.

The re-contextualisation of institutionally-produced contents within the practice of gifting and the reciprocity that this assumes confers on these contents a symbolic value as a part of the very cultural capital defining both group participation and the group itself. In addition, the value placed on these practices for young people entail the potential for them to show some resistance within those dynamics of family negotiations that shape access to the PC as either (a) a platform for archiving content, (b) a tool to archive contents on DVD and DivX for giving as gifts and exchanging and (c) as a platform for re-distributing content via the internet.

Another limitation in this process is the (real of imagined) levels of literacy as regards the skills to manage audiovisual material within this distribution network. Managing audiovisual content is supposed to be complicated and difficult (a supposition that is shared by youth when learning form their peers, given that usually such skills do not arise simply from direct experience of the technologies). Meanwhile, mobile devices are still perceived as being too expensive and at the moment as lacking sufficient technological support.

Conclusions

As the research data show, consumption among young Italians (representing, potentially, a cutting edge for broader tendencies within society) has to be conceptualised as a strongly interconnected set of practices. Young people are increasingly able to switch between different technological platforms and different contents (both those created by the media companies and user generated contents). And they are engaged in redefining their relationship with these media, both in terms of the social role played by media and the technologies, places, times, patterns and rituals of consumption practices. Henry Jenkins is therefore undoubtedly right when, discussing the new media landscape, he describes it as a stratified, often contradictory ensemble of consumption practices. It is certainly possible to 'describe such a scenario in terms of convergence, but steering clear from the usage the term has usually acquired in media and technology circles: the utopian dream that today's chaotic and often redundant array of communication technologies will someday co-alesce into an elegant and all encompassing singularity, a monolithic medium for every kind of message' (Sinnreich, 2007, p.44). Instead, what we are witnessing is an ever-increasing complexity in the possible ways of consuming media. These offer themselves, to use an expression dear to the field of internet studies, as a hypertext of possible pathways, as an ensemble of possibilities for use, where consumption is based upon different variables that influence the relative choices. These pathways are, as we demonstrated, individual in nature and often idiosyncratic in their singularity. However, we still can map them according to shared patterns and highlight those common aspects that are rooted in the generational and national specificities of the youth analysed here.

We would like here to comment upon two aspects that seem to characterise young Italians' cross-media consumption patterns.

The first element is the centrality bestowed by this group upon the relational element of communication, i.e. communication with social networks. As research by Mediappro (2006) (albeit limited to adolescents and pre-adolescents) also indicates, we are dealing here with the centrality, in our country, of media and technologies that sustain interpersonal relations (such as the mobile phone in the past few years and currently instant-messaging software). This contrasts with some other countries where media production and social participation play more relevant roles in young people's relationship with the digital landscape.

The attribution of value to user-generated content, to amusing videos and even to file-sharing, is in this sense emblematic. These are often conceptualised as resources primarily in relation to their social spendability, at one moment in time having a purely phatic function (e.g. when showing or sending amusing videos to show you are thinking of others), at another moment acting as resources for developing or expressing a sense of identity (e.g. user-generated content) and at yet another moment serving as a tool for the consolidation of relationships (e.g. file-sharing withing the peer group).

The second, particularly Italian, dimension is related to the particular history of our media system, dominated up until the last few years by the centrality of general commercial television. Digital TV and new media have definitely and crucially broken this dominance. But older media, at the very least, retain a strong symbolic importance, as witnessed by the role still played by mainstream media in orienting the cross-media consumptions of young Italians.

Young Italians are part of the so-called 'web 2.0 generation' or 'ipod generation' and are actively participating in the processes of transnational media consumption. Nonetheless, particular social variables such as age or nationality, as well as the ways in which media have been socialised and ICTs incorporated within different national and cultural contexts, continue to play an integral role in shaping of the processes by which media technologies are integrated within the framework of convergence culture.

Bibliography

Aroldi, P., et al., 2006. Digital terrestrial television in Italy: approaching the audiences. In: N. Leandros, ed. 2006. *The impact of the internet on the mass media in Europe.* Suffolk (UK): Abramis, pp. 463-475.

Benkler, Y., 2006. *The wealth of networks: how social production transforms market and freedom.* New Haven: Yale University Press.

Bennato, D., ed. 2007. *I comportamenti di consumo di contenuti digitali in Italia: i caso del file sharing, rapporto di ricerca.* Rome: Fondazione Luigi Einaudi per studi di Politica ed Economia.

Colombo, F. & Vittadini, N., eds. 2006. *Digitising TV: theoretical issues and comparative studies across Europe.* Milan: Vita e Pensiero.

Gitelman, L., 2006. *Always already new: media, history and the data of culture.* Cambridge (MA): MIT Press.

Gitelman, L., & Pingree, G.B., 2003. *New media, 1740-1915.* Cambridge (MA): MIT Press.

Haddon, L., 2005. Research Questions for the evolving communications landscape. In: R. Ling & E. Pedersen, eds. *Mobile communication: re-negotiation of the social sphere.* London: Springer-Verlag, pp. 7-22.

Haddon, L. et al., eds. 2005. *Everyday innovators: researching the role of users in shaping ICTs.* Dordrecht: Springer.

Hartmann, M., 2006. A mobile ethnographic view on (mobile) media usage? In: J.R. Höflich & M. Hartmann, eds. 2006. *Mobile communication in everyday life: ethnographic views, observations and reflections.* Berlin: Frank & Timme, pp. 273-298.

Hine, C., 2000. *Virtual ethnography.* London: Sage.

Ito, M., 2007. Networked publics. Introduction. In: K. Varnelis, ed. 2008. *Networked publics.* Cambridge (MA): MIT Press, pp. 1-14.

Jenkins, H., 2006. *Convergence culture: where old and new media collide.* New York: New York University Press.

Jenkins, H. & Thornburn D., 2003a. *Democracy and new media.* Cambridge

(MA): MIT Press.
Jenkins, H. & Thornburn, D., 2003b. *Rethinking media change: the aesthetic of transition*, Cambridge (MA): MIT Press.
Lievrouw, L. & Livingstone, S., eds. 2002. *Handbook of new media: social shaping and consequences of ICTs*. London: Sage.
Livingstone, S., 2002. *Young people and new media: childhood and the changing media environment*. London: Sage.
Marcus, G., 1995. Ethnography in/of the world system: the emergence of multi-sited ethnography. *Annual Review of Anthropology*, 24, pp. 95-117.
Mediappro, 2006. *A European research project: the appropriation of new media by youth*. [Online] Available at: http://www.mediappro.org/publications/_finalreport.pdf. [Accessed 9 February 2010].
Nielsen//NetRatings, 2007. *Web 2.0*. [Online] Available at: http://www.nielsen-online.com/pr/PR_040407_IT.pdf [Accessed 9 February 2010].
Pasquali, F., 2003. *I nuovi media. Tecnologie e discorsi sociali*. Rome: Carocci.
Scaglioni, M. & Sfardini, A., 2007. *MultiTV*. Rome: Carocci.
Scifo, B., 2005. *Culture mobili. Ricerche sull'adozione giovanile della telefonia cellulare*. Milan: Vita e Pensiero.
Silverstone, R. Hirsch, E. & Morley, D., 1992. Information and communication technologies and the moral economy of the household. In: R. Silverstone & E. Hirsch, eds. *Consuming technologies: media and information in domestic spaces*. London: Routledge, pp. 15-31.
Sinnreich, A., 2007. Come together, right now. *International Journal of Communication*. [Online] 1, p. 44. Available at: http://ijoc.org/ojs_index.php/ijoc/article/view/48/12. [Accessed 9 February 2010].

Frank Thomas, Nicoletta Vittadini and Pedro Gómez-Fernández

Chapter three. Cultural influences on the adoption of web 2.0 services

Introduction

"It is the broadband services that are likely to have the most pervasive effect on our lives and culture…" (Cutler & Company Pty Ltd, 1994).

"It's not about the technology or the artefact, but about the culture in which those technologies and artefacts are embedded" (boyd, 2007).

The roll-out of broadband infrastructure and the socially inclusive use of broadband-based services are central issues in Europe's path towards an information society. The way in which broadband infrastructure itself diffuses including the availability of broadband connections, connection speed and tariff structures, discussed in chapter 5 of this book) has important consequences. However, the actual take up of broadband services is also clearly significant, even if, according to recent studies, broadband diffusion and service adoption in Europe are driven more by supply than by demand (Crabtree, 2003). One thing we need to know about the adoption that does take place is whether there is a common model for all European countries that helps to us to understand broadband take up and use or whether we have to take national differences into account.

This chapter addresses this issue and is based on two major hypotheses. The first is that a major under-researched influence on broadband adoption might lie in the cultural differences between countries. Besides basic socio-economic and socio-demographic influences, we already know that major cultural variables can affect broadband diffusion and service adoption, through such factors as national consumer perceptions of technology, perceptions of risk and considerations about how broadband use might relate to social capital (Heres & Thomas, 2007). Moreover, these variables influence not only the speed of broadband adoption, but also the kind of devices and services used.

The second hypothesis is that among all broadband services the most significant ones, revealing new user behaviour and cultural differences among European countries, are web 2.0 services (defined in more detail below, but including many audio-visual elements and services requiring more user activity compared to looking up information on the web). There are several reasons for this. These services are the most dependent on broadband diffusion (in that many cannot be used effectively without a broadband connection). Their use can be considered to be an indicator of the development of the information society as they require specific technological skills and are also a marker of the diffusion of digital literacy (Scott & Julian, 2008). This is important because the degree to which citizens can participate in the information society potentially expands the field of creative opportunity and increases individual involvement in cultural, economic and political life (Benkler, 2006). Lastly, current web 2.0 service diffusion is entering its mature phase, which followed a period of exponential growth, where it is now shifting from global to more national and niche applications (Pascu, 2008).

Besides the traditional innovation and diffusion models that already included social values (Rogers, 1995; Mackay & Gillespie, 1992), domestication analysis provides an appropriate theoretical framework for interpreting the adoption of broadband services in different European countries. This is because it can take into account the cultural factors affecting the appropriation of technologies in domestic cultures, in accordance with the household's values and interests (Silverstone & Hirsch, 1992). In the following sections we shall first provide an overview of the cultural variables potentially affecting broadband adoption and usage. We will then explain our analytical model and it is operationalised in the following analysis, describe the available data sources and present the results of the multivariate analysis of these data. We shall conclude with some ideas about deficiencies in the current state of understanding of broadband and web 2.0 adoption and usage.

Broadband studies and cultural variables

Exploratory research on the adoption and diffusion of broadband started in 2003/2004. One of the first conceptual models aimed at understanding consumers' take up of broadband was based on household consumption practices. It was developed in order to identify the motivations for adoption of and resistance to innovations and factors that can accelerate the diffusion process (Choudrie, 2004). The variables taken in account were the role of user experience (Oh, 2003), users' socio-demographic profiles (Stanton, 2004) and their socio-economic attitudes (Choudrie, 2004).

More recent survey reports (e.g. the Special Eurobarometer 293 report, TNS Opinion & Social, 2008a) have pointed to additional influences on broadband diffusion in European households, such as the number of household members and household composition, whether users live in urban or rural areas, and the number of computers in households.

> "Households with several members are significantly more likely to have a computer than single households…there has been a strong increase in computers in multiple member

households in the NMS12, while computer possession in single households seems to have stabilised at a relatively low level." (TNS Opinion & Social 2008a, p.51).

The level of diffusion of the narrowband internet that preceded broadband has also proved to be a factor.

"The higher the overall Internet penetration rate, the higher the share of broadband connections in comparison to narrowband connections." (TNS Opinion & Social 2008a, p. 56).

This helps us to understand why broadband diffusion is more advanced in the older member states of the European Union[24].

Potential 'cultural' influences on broadband diffusions have received less attention. In fact, what counts as 'cultural' is the subject of some debate since it is one of the most contested notions in the social sciences (Livingstone, 2003; Haddon, 2005). However, there have been efforts to identify variable potential cultural influence, where an operational definition might include social values, beliefs and attitudes, and their translation into everyday social perceptions, behaviour, and material artefacts (see chapter one, the expanded version of which is Thomas, et al., 2005). In other words, 'culture' covers commonly shared symbols and a broad range of meanings and representations of the world. It includes behaviour patterns as embodied over time in habits and customs as well as the way they are expressed in material objects such as the styles and organisation of places.

As already noted, this study assumes that cultural variables influence not only broadband adoption, but in particular the adoption of web 2.0 services. This is based on the fact that Rogers' diffusion of communications and media innovations (Rogers, 1995), the Technology Acceptance Model used to study management information systems (Davis, 1985) and its extension to the study of e-mail perception and use (Straub, 1997) all suggest that people have different levels of acceptance and use not only of a new technology but also of new services and new types of content.

We will came back to this issue in a later section on methodology, but before describing the specific cultural variables employed in this analysis it is useful to describe the other object of research – web 2.0 services – and their relevance.

---

24 "Broadband technology is clearly less prevalent in the new Member States than in the old ones, where just under a quarter of households have broadband access." (TNS Opinion & Social 2008a, p. 56).

Broadband and web 2.0 services

Broadband not only permits higher connection speed, but it also changes the way that the internet is used in domestic spaces. For instance, it liberates users from the plain old telephone service so that conventional voice services are not interrupted by internet access. It also allows users to be connected all the time (the 'always on' feature), to have fast and simple logins, and to share network access across several computers (Scott & Julian, 2008). Using a broadband connection to access web 2.0 services can modify household activities and the routines of media consumption, in turn increasing the importance of the internet connection in households. For some, broadband can lead to a reduction in TV watching and newspaper reading. It can lead to an increase in website creation, blogging and the posting of commentaries on other people's blogs. It can result in more downloading, listening of online television and radio, sharing content among friends. And finally it can encourage some users to do more social tagging.

Web 2.0 is, in fact, an umbrella term so large that it is not so useful to take it as the unit of analysis. Hence, for analytical reasons, web 2.0 has been broken down into groups of services that share the fact that they have comparable user practices. This will enable us to continue with a more fine-grained analysis that compares the take-up of these different types of web 2.0 services in different countries. We thus identified three types, or groups, of services (allowing for some overlap):

– *Creative internet*: 'creative uses' of the net ranging from relatively straightforward user-generated content such as sharing photographs to the distribution of more complex amateur-produced material" (Scott & Julian, 2008).
– *Social Computing*: 'social uses' of the net including collaboration, sharing and communication (blogging, podcasting, uses of Wiki applications, social networking, multimedia sharing, social tagging and social gaming) (Pascu, 2008; Osimo, 2008).
– *Circular entertainment:* 'consumer use' of the net ranging from the use of peer-to-peer networks to acquire cultural products to downloading, podcasting and streaming. Much of this consumption, rather than production, involves entertainment products that circulate amongst internet users.

Cultural variables and broadband and web 2.0 adoption

The identification of the cultural variables to be used in the analysis is in part based on a comprehensive list of cultural variables affecting ICT adoption in general that was outlined in chapter one. These potential cultural influences included various aspects of social structures (covering, for example, social homogeneity and stratification, religion, education and literacy, mass media, language and the patterns of international communication). The included time structures (i.e. subjective experience of time, patterns of time use and expectations about, for example, punctuality). Other cultural influences related to social values (in particular openness to technological innovation, and the degree of individualism), the nature of communication (communication forms, patterns and expectations about how to communicate), and material culture (for example, spatial layout of the settlement system, housing, artefacts).

In addition, we considered socio-demographic variables that have been shown to influence adoption: the mean size of the household (larger households are more inclined to get internet access than smaller ones) and residential stability (which supports the maintenance of existing social contacts). In addition, in the present analysis we considered several other variables that have emerged specifically from studies of the relationships between country differences and internet adoption. They are not all 'cultural' influences per se, but they have been proven to affect internet use.

The first of these extra variables is *interpersonal trust*[25]. At the household level, studies examining both US and European data found a significant correlation between internet information exchange, interpersonal trust and civic participation after controlling for socio-demographic influences (Shah, 2001a, 2001b; Heres & Thomas, 2007). In addition, the link between internet diffusion and interpersonal trust at an aggregate level has also been confirmed in various global cross-country studies (Bornschier, 2001; Volken, 2002; Huang, Keser, Leland & Shachat, 2002). More specifically, Cho confirmed the relationship between the aim to be socially connected, the frequency of using email and the internet and the level of interpersonal trust (Cho, 2003). Lastly, the relation between internet use and interpersonal trust has been successfully explored utilising both the uses and gratifications framework and the cognitive mediation model[26] (Beaudoin, 2008).

The second variable, *social capital,* has been consistently found to affect internet adoption and use no matter which definition of social capital had been used and after standard controls for socio-economic influences at the individual level had been introduced (Franzen, 2003; Ling 2007; Thomas 2008). For

---

25 The definition of trust used is derived from the sociological and economic literature on trust saying that the existence of uncertainty between the agents involved is a crucial factor present in most definitions of trust (Dutton, & Shepherd, 2003). In order for trust to exist, there must also be risk. Without trust, risk is paralysing; interaction simply does not take place. If the level of trust surpasses the threshold of perceived risk, then the more trusting people will engage in risk-taking in relationship. Trust, then, enables action in the face of risk allowing for uncertainty in risky situations (Luhmann, 2000).

26 The cognitive mediation model posits that the motivation to use media, actual media use, and information processing interrelate in predicting knowledge (Eveland, 2001).

example, one major use of the internet is for emails for interpersonal communication, and socially more connected people use the internet more often than less well connected ones, both for making formal as well as for making informal contacts. As regards how the concept is operationalised in studies, social capital has been measured in terms of, for example, the level of volunteering (for bridging social capital) and in terms of meeting with friends (for bonding social capital).

The third set of variables considered here is composed of diverse influences that have been identified as being *cultural variables*. Their selection is based on the fact that previous studies of web 2.0 service take up have underlined how differences in country-specific ways of adopting may be retraced to certain types of national differences, for instance linking differences between national blogospheres with differences in the rules of social relations (Schlobinski & Siever, 2005). One of the most widely used groups of these cultural variables is the one developed by the Dutch social psychologist G. Hofstede. He tried to understand the organisational cultures of countries through five basic patterns that he calls power distance, uncertainty avoidance, collectivism, masculinity, and short-term versus long-term orientation (Hofstede & Hofstede, 2005). Hofstede defines power distance as the social inequality accepted by the powerless, uncertainty avoidance as a preference for avoiding risk through norm orientation, collectivism as the extent to which an individual is integrated in larger groups, and masculinity as an indicator of assertive and competitive behaviour. Time orientation has not been taken into account as it had only been introduced by Hofstede to distinguish Asian countries.

Hofstede's approach has been criticised on the grounds that his fieldwork was based on relations in the workplace and one cannot use it to study a country's culture because the original survey that generated these categories was conducted inside the national branches of the transnational business firm IBM (McSweeney, 2002; Erumban & de Jong, 2006). However, his categories have been used here because several cross-country studies of ICT adoption seem to have confirmed its basic tenets (Hofstede & Hofstede, 2005). For example, ICTs have been more easily adopted in countries with low power distance (Hasan & Ditsa 1999; Bagchi, et al., 2004) or low uncertainty avoidance (Hasan & Ditsa, 1999; Erumban & Jong 2006). The influence of high individualism and of uncertainty avoidance have been confirmed by a cross-country study of mobile phone diffusion (Sundqvist Frank and Puumalainen 2005), by a study of trust in online bidding (Vishwanath 2004) and by a meta-analysis of technology diffusion (Cardon & Marshall 2008). To take a particular example in more detail, the influence of Hofstede's cultural traits was even reflected in different styles of contributing to Wikipedia. Pfeil and her co-authors analysed the inter-action between individualism and the frequency of corrections and additions to Wikipedia in more and less collectivist countries. In those with a low level of individualism Wikipedia voices were more often corrected and there were more frequent additions. These researchers also demonstrated that the cultural differences that existed offline also existed online using Hofstede's indicators (Pfeil, et al., 2006).

These studies have focused more on participation in web 2.0 communities than on the adoption of specific services. Nonetheless they have the advantage of stressing the relevance of cultural factors for user actions (e.g. individual participation in social networks, uploading content, etc.) and they have already been used in web studies[27]. The potential relevance of Hofstede's cultural variables can be shown by two further examples. Power distance is greater when people are more willing to accept an unequal society. As web 2.0 services essentially further non-hierarchical social exchanges it is therefore logical that a society with a high power distance should be slower in adopting participative forms of the web. Meanwhile, uncertainty avoidance characterises societies that favour rule orientation instead of case by case decisions. In principle, the interactive web, with its unpredictable encounters and behaviour, should therefore be more easily adopted in cultures with low uncertainty avoidance.

Another potential cultural influence on broadband use can be post-materialistic values. Post-materialism is a theory about societal value change developed by Inglehart (1977, 1990). He saw a set of values emerge in modern societies after long periods of material affluence that lay the stress less on fulfilling material needs, but on attaining more abstract values, such as happiness, the protection of the environment, and self-fulfilment. This can translate into a social participation, such as in the form of self-help groups. Norris, for example found some support for an effect of post-materialism on cyberculture (Norris, 2001). Meanwhile, Choi, et al. (2004) confirmed the positive effect of valuing self-fulfilment upon internet usage. Two further values or societal goals – demanding that your society be innovative and tolerant and accepting free speech – have been added to the analysis as separate influences due to their importance for a free development of the internet. The symbolic integration of internet users, i.e. their identification with a large community of innovative users that are felt to be in advance of their time, was included since this can make users perceive non-users as being old-fashioned, a typical attitude shown towards late adopters (Rogers, 1995).

Finally, among the social behaviours that allows to characterise a culture and that are known to influence internet usage are what we might call 'cultural activities', in particular among youth (Cardon & Granjon, 2003). Cultural activities can be either consumptive – for instance, going to the cinema – or

---

27 These dimensions have been found to be relevant to web design (Marcus & Gould, 2000; Robbins & Stylianou, 2002; Sheridan, 2001) and aspects of web-based communication (Tsikriktsis, 2002; Wilson, et al., 2002).

active – producing a video, singing, acting etc. Meanwhile, having had a long experience in living every day with the internet has been shown to be another behaviour that facilitates the adoption of web 2.0 services. Therefore both sets of influences have been integrated into the indicators that will profile a country's culture.

The list of potential influences on internet and web 2.0 usage outlined in this section have been integrated to form a list of indicators shown in Table 1.

The data used

The analysis conducted here is based on data aggregated on a country level from the European Social Survey (ESS Round 1, 2002), the Eurobarometer and Flash Eurobarometer surveys, the International Telecommunication Union and Hofstede (Hofstede & Hofstede, 2005). The quality of the surveys at hand used varies, the best data coming from the European Social Survey. Since we are studying the potential influence of culture we do not analyse the action of individual users but rather look at the country level, combining information from various aggregate sources. However, the restriction of the analysis to country-level information evens out a potentially important variation at the level of the individual users even if it clearly focuses the study on explaining the collective forces at work. There are additional limits inherent in the data due to sampling. All Eurobarometer data exclude residents with non-European citizenship, which might bias the results (see for instance TNS Opinion & Social, 2008, Technical Specification). Finally, the choice of countries was considerably limited due to the Hofstede data, which only exist for 18 European countries among those that also participate in the Eurobarometer surveys. The detailed survey of cultural activities from Eurobarometer 56.0 conducted in 2001 (Spadaro, 2002) was not used as its information regarding the internet use was deemed to be out of date. The list of surveys can be found in Table 1.

Variables	Flash Eurobarometer 241, 2008	Special Eurobarometer 278, 2007	European Social Survey, 2002	Hofstede & Hofstede 2005	Eurobarometer 69.1 2008	Eurostat 2005	International Telecommunications Union
Values and attitudes	Value of symbolic integration of internet (non users perceived as old-fashioned)	Societal goals: Progress and innovation; Freedom of opinion; Tolerance and openness to others (Values intended to preserve and reinforce society); Localism (local attachment to village, town, city)	Social trust; Family orientation (Importance of family life more important than friends); Openness to friendship (contacts with friends from abroad)	Power distance (accepted social inequality); Uncertainty avoidance (openness to change)	Postmaterialism (orientation to participation and freedom vs to stability and control)		
Social behaviour		Off-line cultural performance and production rate (presence of cultural performance activities -sing, act, dance - and of cultural production activities - photography, film and visual arts); Cultural consumption (mean frequency of cultural consumption, breadth of cultural consumptions)	Social capital: formal (volunteering rate); non formal (frequency of meetings with other people)				Long-term experience in digital life (internet users in 2000, per cap.)
Socio demographic variables			Orientation of communication channels (size of household); Sedentariness (length of residence)			Leisure capacity (Hours worked per week.)	

Table 1: List of indicators used in factor analysis and their sources

Table 2 shows indicators taken from two representative European surveys that the study used in order to describe the three main categories of web 2.0 services: social computing, creative computing and circular entertainment.

Web 2.0 services	Source: Flash Eurobarometer 241, 2008	Source: Special Eurobarometer 278, European Cultural Values 2007
Social computing	Creating a profile or sending a message to a social network	Send/receive pictures Exchange files
Creative Internet	Uploading photos or videos Creating own website or blog	Create own website or blog Visit blogs
Circular Entertainment	Playing or downloading games, music, film, software. Transferring contents from internet to personal devices	Play games. Download free music Download free movies, TV Listen to radio, music Watching TV

Table 2: A typology of web 2.0 services and their survey sources.

Key dimensions for differentiating countries

The data analysis has several aims. The first one is to identify a set of country types based on the socio-cultural variables already mentioned. The second is to examine whether the kind of web 2.0 services adopted is linked to these country types.

In principle, in order to describe a culture in statistical terms a wide array of indicators has to be used. However, in practice it is too difficult to analyse the variety of influences on adoption without consolidating these factors into fewer variables. Hence, we need to reduce the array of influences into a few scales or dimensions that capture all those variables at work. Then we can show how different countries are positioned on the most important, overarching dimensions. Factor analysis is a standard tool used for achieving this. It takes the data relating to the variables, looks at the vast number of interactions between different variables, and indicates the patterns where the influence of variables correlate, in order to consolidate these down into as few scales as possible, known as 'factors'. The standard procedure is than to give those factors names that capture the essence of those interactions, and describe them, after which one can locate countries on scales based on these factors.

The factor analysis we conducted resulted in the following three dimensions. The factors were then characterised below in terms of the strongest influences within them (for details, see Table 3).

Factor 1: Participation & culture activity Countries high in terms of this factor strongly value social capital (through volunteering) and open relations with foreigners, which in turn translates into a remarkable level of interpersonal trust[28]. They have a penchant for post-materialistic values, i.e. seek individual self-fulfilment in life coupled with a tendency for tolerance. They are not afraid of uncertainty. They love all approaches to culture: as amateur artists people in these countries produce cultural content, but they also consume a large variety of cultural offers[29]. Their populations have had a long experience with the internet. Their time structure allows for abundant leisure, in part due to the absence of children, suggested by the average small size of their households. Their geographical mobility translates into a low local attachment[30]. All of these features together make them open to technological innovation and engagement with the participative web.

Influences	Factor scores :		
	Participation and culture	Technology and sociability	Egalitarian society and active culture
Postmaterialism	0.59	0.30	0.42
Non-users perceived as old fashioned	0.44	0.53	0.36
Tolerance as a global value	0.35	0.26	
Interpersonal trust	0.32	0.37	0.61
Uncertainty avoidance	-0.32	0.25	0.84
Power distance			0.94
Free opinion as a global value		0.36	0.28
Visual & manual artistic activities	0.80	0,23	0.30

---

28 Post-materialistic values can be understood as furthering individual self-fulfillment. See the list of indicators used in factor analysis in annex 2.
29 For instance, they like to sing as well as to go to a musical concert, to take pictures and to visit exhibitions.
30 For a complete list of the variables and their sources see annex 1.

	Factor scores :		
Frequency of Cultural consumption	0.69	0.26	0.52
Volunteering	0.66	0.30	0.37
Width of Cultural consumption	0.64		0.66
Traditional artistic activities	0.61	0.24	0.55
Friends from abroad, in Europe	0.41	0.34	0.33
Friends from outside Europe	0.34	0.54	0.24
Relative importance of family vs. friends	-0.35	0.69	0.35
Local attachment	-0.23	0.23	
Frequency of socialising		0.72	0.33
Household size	-0.77		
Hours worked per week	-0.70	0.40	
% Internet users in 2000	0.46	0.63	0.46
Length of residence	-0.26	0.86	0.21
Explained variance	53%	13%	8%

Table 3: Rotated factor matrix Scores between -.20 and .20 have been eliminated to increase readability.

Factor 2: Technology & sociability This factor describes countries that particularly value friends more than family when constructing new networks after people have moved home frequently. Their long internet experience and an inexpensive internet go together with an interest both in social integration – through social networks allowing people to stay in touch – and in symbolic integration – through the importance attributed to technological innovation and the perception of non-users as being marginal and out of date. Their populations generally love artistic activities, but less than the populations in the other groups of countries. Their social openness also shows up in their sociability with friends coming from foreign countries and in strongly favouring free speech.

Factor 3: Egalitarian values & active culture The populations of these countries strongly value a society with a flat hierarchy and weak rule-orientation, which in turn translates into an openness to change. They place a high value on friendship and interpersonal trust and are involved in performative and productive cultural activities.

Grouping countries

The first factor explains a considerable 53% of the statistical variation, the following two 13% and 8% respectively[31] The next stage of the analysis involved identifying how countries were located in relation to these factors, specifically in relation to scales that measure the most important dimensions, and using this as a basis for grouping them. Hence, the first two factors identified above were then visually cross-tabulated to create typical combinations of countries based on their socio-cultural structure, where participation and culture activity forms the horizontal axis in the Figure 1 below and the technology and sociability dimension forms the vertical axis.

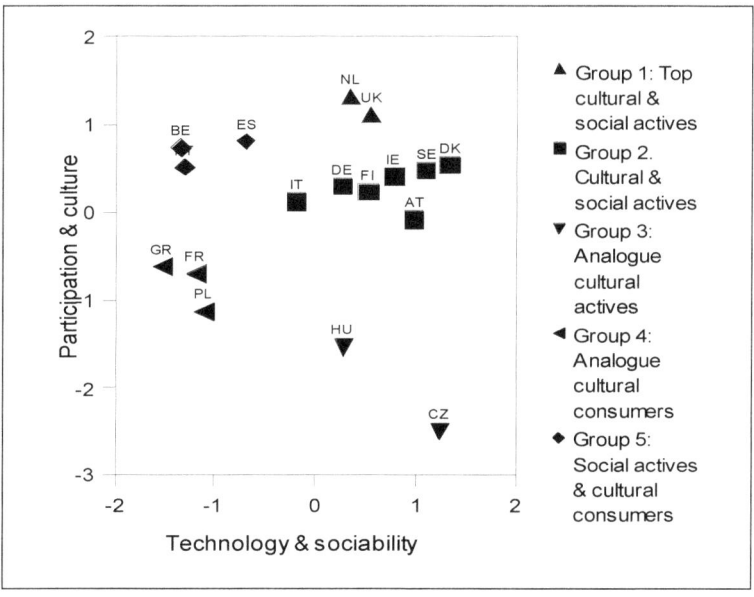

Figure 1: Cultural country types and their components

The five cultural country types found through this process were labelled based on the dominant features of their underlying factors. They are, starting in the upper right corner of Figure 1:

---

31 The remaining factors were not interpreted as they only have a limited influence.

- *Group 1:* The two countries, the Netherlands and the UK, are characterised by a strong social and cultural activity and the accompanying values of self-fulfilment, openness and technological innovation. Both countries show the most favourable conditions for broadband use. We label them the *top culturally and socially active* countries.
- *Group 2:* Denmark, Sweden, Ireland, Austria, Finland, Germany and Italy are close to group 1, but are significantly lower in the figure in terms of technology and sociability. They can be labelled *culturally and socially active* countries. Italy can be considered to be at the centre of all European countries studied.
- *Group 3:* The group puts together Hungary and the Czech Republic, two countries characterised by social behaviour and values linked to the first factor (Participation & culture activity), with an intense social participation, interpersonal relations, and cultural production diffused also at the individual level, but weaker in technological innovation as a tool to improve social life and in integration and social openness, too. We label them *analogue cultural active* countries.
- *Group 4:* The countries in this group, Greece, France and Poland, are characterised by both weak activity in terms of participation & culture activity, a high value attributed to interpersonal relations and cultural consumption (not production at the individual level) and being low in terms of technological innovation as a tool to improve social life and to facilitate integration and social openness. Therefore, we label them *analogue cultural consumers*.
- *Group 5:* This group is composed of Belgium, Spain, and Portugal, countries characterised by intense social behaviour and social values linked to the factor technology & sociability, but that are weak in terms of cultural activity and higher social participation. We label them *socially active and cultural consumer* countries.

It is noteworthy that the cultural influences studied by Hofstede do not, it seems, exert a great influence. Here we focussed on uncertainty avoidance and power distance, because both influences have been cited several times cited in previous research. The reason for their limited impact will probably be that they are highly intercorrelated ($r=.79$) and that both in turn correlate with postmaterialism ($r=-.66$ and $-.52$) for the European countries studied. So the self-fulfilment aspect of postmaterialism, an important variable for establishing the country typology, might have reduced the effect of Hofstede's variables.

Country groupings and the adoption of web 2.0 services

Starting from these country types the study next tried to verify whether differences in cultural variables are correlated to the diffusion of web 2.0 services and in particular to the different kinds of services classified earlier as circular entertainment, social computing, and creative internet services. The first step involved comparing the diffusion levels of web 2.0 services in the different countries. Then results have been grouped as a country typology (see Table 2).

The diffusion of web 2.0 services among households with an internet connection at home is quite high. The lowest usage level – 33% of users in any country type and in any service – can be found in Belgium, Portugal, and Spain (countries we defined as *social active and cultural consumer* countries) in relation to social computing services, (an example of which is maintaining a personal profile in a social network). The highest levels of diffusion, 62% of users, can be found for circular entertainment, i.e. downloading and gaming, in countries labelled as *analogue cultural consumer* countries. This means that differences between countries are subtle, but meaningful.

Before going into a detailed analysis some more general observations have to be made. You will not always find the strongest concentration of users of a given service in the countries where you might imagine they would be if only looking at internet diffusion. Downloading music, for instance, is most prevalent in those countries where the internet has a low diffusion in general, probably because in these countries adoption is still dominated by younger users who have fully domesticated this new way of distributing music. At the time the data were surveyed, in 2007 and 2008, social networking sites were most popular in Poland and Hungary and less popular in the Latin countries (France, Italy and Spain), i.e. in countries where off-line social activities such as going outdoors to socialise are more appreciated. The publication of visual content online turns out to be most common in the East European countries that have undergone huge social transformations and a young user population, and less in the countries with a long digital experience, such as Finland or Sweden. Transferring content to and from devices is most common in Poland, Greece and Spain. In other words, countries that are less technology-oriented and in which cultural consumption is more appreciated than its (amateur) production become the most advanced in this activity.

Country type/ Activity	Group 1: Top culturally and socially active	Group 2: Culturally and socially active	Group 3: Analogue culturally active	Group 4: Analogue culturally consumer	Group 5: Socially active and cultural consumer
Playing games, downloading music, etc.	46%[32]	45%	59%	62%	49%
Transferring content from the internet to private devices	42%	38%	41%	49%	46%
Uploading photos, videos to a public website	40%	36%	49%	49%	42%
Creating a private website or a blog	9%	11%	8%	9%	9%
Maintaining a profile on a social networking site	36%	36%	37%	36%	33%

Table 4: Web 2.0 activities by country types.
Source: Flash Eurobarometer 241, 2008, calculations by authors.

Simply examining the history of internet diffusion based on economic factors and populations' orientation towards technological innovation cannot explain such results as those listed above, whereas an analysis of the correlations between cultural variables and the forms of social behaviour utilised as indicators for the country typology could provide a step towards understanding such patterns
In preceding Table 4 we see how the different groups of countries have different adoption rates for selected web 2.0 services.

---

32 In the top left of the table, 46 means that 46% of the population in group 1 countries (the Netherlands and the UK) play games online, download music, etc.

Country groupings and adoption patterns of web 2.0 services

The relevance of cultural variables for explaining adoptions patterns of web 2.0 services can be demonstrated further by analysing the different diffusion patterns in the country groupings, this time looking across all three types of web 2.0 service.

Starting with adoption of web 2.0 services classified as sustaining circular entertainment we can observe that the *top culturally and socially active* countries as well as the *culturally and socially active* countries tend to have an even adoption pattern, in that all kinds of web 2.0 services have similar levels of adoption and sustain both social and cultural activities. These countries are characterised by a wide range of cultural and social activities organised in association with strong social rituals, and through volunteering. At the same time they favour technological innovation and participation through social activities in on line and in off-line life. We define this pattern as being an *homogeneous adoption pattern* because of the absence of any dominant kind of service.

In the *socially active and cultural consumer* and *analogue cultural consumer* countries we can observe a web 2.0 adoption pattern involving mainly the acquisition and consumption of cultural products, in particular using circular entertainment services to increase the choice of products available to users. At the same time, social activities seem to be more orientated to offline life. The populations of these countries see technological innovation and social activities as being tools to promote social integration in all of its forms. We define this pattern as being a *cultural goods-oriented adoption pattern.*

In countries labelled as *analogue culturally active* and characterised by a wide range of cultural consumption and production activities and by a strong value placed on social participation (in formal and informal social life), but whose populations have less interest in technological innovation we find a third adoption pattern. The adoption of web 2.0 services is focused both on cultural activities (consumed and produced or uploaded user-generated contents) and on social computing activities aimed at enhancing participation in society. We label this pattern as being a *cultural and social activities-oriented adoption pattern.*

The weight of culture can be seen in the correlation between the adoption of web 2.0 services, social values, and social behaviours correlated with offline social life. Countries in which social activities are understood as a tool to increase social participation are characterised by higher levels of adoption of web 2.0 social computing services. Countries where social activities are understood as a tool to increase social inclusion and integration seem to be less interested in the adoption of services mediating sociability such as social computing: they live their sociability more in formal and informal social life forms outside the web.

Conclusions

The present study intended to explain the evolution of web 2.0 services in a large selection of European countries by considering the role of cultural influences. The contribution of this analysis lies in the fact that although there are a small number of country-based studies of broadband use, they have not yet been put into a comparative framework that allows us to see eventual cultural variations in the development paths of broadband use in Europe.

After providing an overview of the emerging field of cultural web 2.0 studies we put together a list of potential influences on its take up, such as interpersonal trust, indicators for social capital, measures of current cultural consumption and amateur content production, and social values, in particular the cultural influences studied by Hofstede, as well as a small set of control variables. The next step involved organising the web 2.0 services studied into three categories: social computing, creative internet, and circular entertainment. The potential influences were then condensed into a few dimensions through a factor analysis of country-level survey data and the combination of the two major factors produced in this process provided the basis for a culturally based typology of countries. The five country types found were subsequently labelled based on the level of their social and cultural activities, online as well as offline: the five types resulting were *top cultural and social active*, *cultural and social active*, *analogue culturally active*, *analogue cultural consumer*, and *social active and cultural consumers* countries.

The analysis has shown a broad correlation between the five cultural types we identified and the adoption of web 2.0 services.

We found a tentative link between the cultural frames of European countries and adoption patterns, in three adoption patterns which we call a homogeneous, a cultural goods-oriented and an activities-oriented adoption pattern. This clearly hints at a more general structure in the evolution of Europe's broadband society: there is a common basis which provides a certain unity, but the country types we derived also show the diversity of the national cultures.

Finally, when it comes to the specific cultural influences on the evolution of web 2.0 services using Hofstede's measures of power distance and uncertainty avoidance they were helpful in explaining, but did not revolutionise our insight as they are considerably correlated with the self-fulfilment aspect of post-materialism. It is in combination with the diverse forms of social participation that they exercise their full force.

These considerations also open a new topic of discussion and analysis: which aspects of social life, in particular social participation and social inclusion, are promoted by web 2.0 services and by their imagination and use in everyday life?

There are certain limits inherent in this analysis. First, if the cultural context clearly plays a part at the country level, within-country variations, such as those between Sicily and Lombardy, will probably also play a role. Second, an important source of variation, the web 2.0 usage of individual households, will have to be examined in depth in a separate study. Third, the published data do not distinguish between frequent and infrequent web 2.0 users so that the national aggregates that this study was obliged to use might give a biased view of the situation because of the many low intensity users. Another limitation lay in the limited number of European countries that we were able study due to the limitations of countries covered by both the Hofstede and the Eurobarometer data sets. Finally, a more thorough analysis demands that our quantitative study be combined with qualitative studies to get a deeper insight.

Bibliography

Beaudoin, C. E., 2008. Explaining the relationship between internet use and interpersonal trust: taking into account motivation and information overload. *Journal of Computer-mediated Communication,* 13, pp. 550–568.
Benkler, Y., 2006. *The wealth of networks: How social production transforms markets and freedom.* New Haven, CT: Yale University.
Bornschier, V., 2001. Generalisiertes vertrauen und frühe verbreitung der internetnutzung im gesellschaftsvergleich. *Kölner Zeitschrift für Soziologie und Sozialpsychologie* 53 (2), pp. 233–257.
Boyd, D., 2007. Choose your own ethnography: in search of (un)mediated life. *The Society for Social Studies of Science (4S) Annual Conference.* Montreal. [Online] Available at: http://www.danah.org/papers/talks/4S2007.html. [Accessed 9 January 2010].
Cardon, D. & Granjon, F., 2002. Eléments pour une approche des pratiques culturelles par les réseaux de sociabilité. In : D. Olivier, & P. Tolila. eds. *Les publics. Politiques publiques et équipements culturels*, *Volume 2*. Paris: Presses de Sciences Po, pp. 93-108.
Cho, J. D., 2003. Beyond access: the digital divide and internet uses and gratifications. *IT & Society,* 1 (4), pp. 46–72.
Choi, J. Watt, J. Dekkers, A. & Park, S.H., 2004. Motives of internet uses: crosscultural perspective – the US, the Netherlands, and S. Korea. *Annual meeting of the International Communication Association,* New Orleans Sheraton, New Orleans, LA, 27 May 2004: ICA: New Orleans.
Choudrie, J., 2004. Towards a conceptual model of broadband diffusion. *Journal of Computing and Information Technology* 12 (4), pp. 323-338.
Crabtree, J., 2003. *Fat pipes, connected people - rethinking broadband Britain.* London: iSOCIETY.
Cutler & Company Pty Ltd., 1994. *Commerce in content - building Australia's international future in interactive multimedia markets.* Melbourne: Cutler

& Company Pty Ltd. [Online] Available at: http://www.nla.gov.au/misc/ cutler/ cutlercp.html. [Accessed 8 January 2010].

Davis, F. D., 1985. Perceived usefulness, perceived ease of use, and user acceptance of information technology. *MIS Quarterly,* 13 (3), pp. 319-340.

Dutton, W. & Shepherd, A., 2003. *The social dynamics of cyber trust: confidence and risk on the internet.* Oxford: Oxford Internet Institute.

Erumban, A. A. & de Jong, S. B., 2006. Cross-country differences in ICT adoption: a consequence of culture? *Journal of World Business,* 41 (4), pp. 302-314.

Eveland, W. P., 2001. The cognitive mediation model of learning from the news: Evidence from nonelection, off-year election, and presidential election contexts. *Communication Research,* 28 (5), pp. 571–601.

ESS Round 1, 2002. European Social Survey Round 1 Data. Data file edition 6.1. Bergen: Norwegian Social Science Data Services.

Franzen, A., 2003. Social capital and the internet: evidence from Swiss panel data. *Kyklos,* 56, pp. 341-360.

Gallup Institute, 2008. *Flash Eurobarometer 241. Information society as seen by EU citizens. Analytical report.* Brussels: European Commission.

Garrido, N. & Marina, A., 2008. Exploring trust on internet: the Spanish case. *Observatorio (OBS*) Journal,* 6, pp. 223-244. [Online] Available at: http://www.obs.obercom.pt/index.php/obs/article/view/213/192 [Accessed 30 January 2010].

Haddon, L., 2005. *Cross cultural differences and cultures of research.* Brussels: Cost Office.

Heres, J. & Thomas, F., 2007. Civic participation and ICTs. In: B. Anderson B. Malcolm G. Jonathan & R. Yoel, eds. *Information and communication technologies in society.* York: Routledge, pp. 175-188.

Hofstede, G. & Hofstede, J. G., 2005. *Cultures and organizations—software of the mind.* New York : McGraw-Hill.

Huang, H. Keser, C. J, Leland, J. & Shachat, J., 2002. Trust, the internet, and the digital divide. *IBM Systems Journal,* 42 (3), pp. 507-518.

Ling, R., 2007. Informal social capital and ICTs. In: B. Anderson B. Malcolm G. Jonathan & R. Yoel, eds., *Information and communication technologies in society.* New York : Routledge, pp. 150-162

Livingstone, S., 2003. On the challenges of cross-national comparative media research. *European Journal of Communication,* 18 (4), pp. 477–500.

Luhmann, N., 2000. *Vertrauen. Ein Mechanismus der Reduktion sozialer Komplexität.* Stuttgart: UTB.

Mackay, H. & Gillespie, G., 1992. Extending the social shaping of technology approach: ideology and appropriation. *Social Studies of Science,* 22 (4), pp. 685-716.

Marcus, A., & Gould, E. W., 2000. Crosscurrents: cultural dimensions and global web user-interface design. *Interactions*, 7 (4), pp. 32–46.

McSweeney, B., 2002. Hofstede's model of national cultural differences and their consequences: A triumph of faith. A failure of analysis. *Human Relations,* 55. pp. 89-118.

Norris, P., 2001. *Digital divide: civic engagement, information poverty, and the internet worldwide.* Cambridge, MA: Cambridge University Press.

Oh, S. A., 2003. Adoption of broadband internet in Korea: the role of experience in building attitudes. *Journal of Information Technology,* 18, pp. 267-280.

Osimo, D., 2008. *Web 2.0 for government: why and how?* Seville: IPTS.

Pascu, C., 2008. *An empirical analysis of the creation, use and adoption of social computing.* Brussels: JRC European Commission.

Pfeil, U. Panayiotis, Z. & Ang, C. S., 2006. Cultural differences in collaborative authoring of Wikipedia. *Journal of Computer-mediated Communication,* 12, pp. 88–113.

Rasänen, P., 2005. Is the internet used for interpersonal communciation? A comparison of 15 European Countries. In: T. Toivonen H. Leena & V. Taru, eds. *National European global research seminars of economic sociology 2004.* Turku: Turun kauppakorkeakolu, pp. 159-183.

Robbins, S. S. & Stylianou, A. C., 2002. A study of cultural differences in global in corporate websites. *Journal of Computer Information Systems,* 42, pp. 3–9.

Rogers, E., 1995. *Diffusion of innovations.* New York: The Free Press.

Schlobinski, P. &. Siever, T., eds., 2005. *Sprachliche und textuelle aspekte in weblogs. Ein internationales projekt. [Linguistic and textual aspects of weblogs].* Hannover: Networx, 46.

Scott, E. & Julian, T., 2008. Broadband and the 'creative internet'. Australians as consumers and producers of cultural content online. *Observatorio (OBS*) Journal,* 6, pp.187-208. [Online] Available at: http://www.obs.obercom.pt/index.php/obs/article/view/215/190 [Accessed 9 January 2010].

Shah, D. V., 2001a. 'Connecting' and 'disconnecting' with civic life: patterns of internet use and the production of social capital. *Political Communication*, (18), pp. 141–162.

Shah, D. V., 2001b. Communication, context, and community: an exploration of print, broadcast, and internet influences. *Communication Research,* 24 (4), pp. 464–506.

Sheridan, E. F., 2001. Cross-cultural web site design: considerations for developing and strategies for validating locale appropriate on-line content. *MultiLingual Computing & Technology*, 12 (7). [Online] Available at: http://www.multilingual.com/articleDetail.php?id=595 [Accessed 9 January 2010].

Silverstone, R. & Hirsch, E., eds. 1992. *Consuming technologies. Media and information in domestic spaces.* London: Routledge.

Spadaro, R., 2002. *La participation des Européens aux activités culturelles. [Europeans' participation in cultural activities]* Brussels: European Commission.

Stanton, L., 2004. Factors influencing the adoption of residential broadband connections to internet. *The 37th Hawaii International Conference on System Sciences (HICSS'04), Track 5.* Big Island, Hawaii, USA 5-8 January 2004.

Straub, D. W., 1997. Testing the technology acceptance model across cultures: a three country study. *Information and management,* 33 (1), pp. 1-11.

Sundqvist, S. Lauri, F. & Puumalainen, K., 2005. Cross-cultural adoption of wireless communications: effects of cultural distance and country characteristics. *Journal of Business Research* 58 (1), pp. 107-110.

Thomas, F. Haddon, L. Gilligan, R. Heinzmann, P. & De Gournay, C., 2005. Cultural factors shaping the experience of ICTs: an exploratory review. In: L. Haddon, ed. *International collaborative research. Cross-cultural differences and cultures of research.* Brussels: COST, pp. 13-50

Thomas, F., 2008. The social capital of migrants and individual ICT use. A comparative analysis of European countries. *The Good, the Bad and the Unexpected.* Moscow 23-25 May 2007. COST: Brussels, pp. 775–795.

TNS Opinion & Social, 2007. *Special Eurobarometer 278. European cultural values.* Brussels: European Commission.

TNS Opinion & Social, 2008a. *Special Eurobarometer 293. E-communications household survey.* Brussels: European Commission.

TNS Opinion & Social, 2008b. *Eurobarometer 69.1 Values of Europeans.* Brussels: European Commission.

Tsikriktsis, N., 2002. Does culture influence web site quality expectations? *Journal of Service Research,* 5 (2), pp. 101-112.

Volken, T., 2002. Elements of trust. The cultural dimension of internet diffusion revisited. *Electronic Journal of Sociology,* 6 (4). [Online] Available at: http://www.sociology.org/content/vol006.004/volken.html [Accessed 9 January 2010].

Vishwanath, A., 2003. Comparing online information effects: a cross-cultural comparison of online information and uncertainty avoidance. *Communication Research* 30 (6), pp. 579-598.

Robert Pinter, Annika Bergström, Fruzsina Gyenes, Leslie Haddon and Francesca Pasquali

Chapter four. Cross-cultural differences in press framing of the internet

Introduction[33]

Although the internet is finding its way into more and more European households there are still considerable national differences in terms of take-up generally, in the adoption of newer developments like broadband, in overall amount of usage and, indeed, in consumer interest in particular services – as illustrated below in table 1 for the four countries taking part in this media analysis project.

There will undoubtedly be many factors contributing to the variation shown below. Some influences on these figures come from the supply side – for example, to what extent access is actually available. Factors here include access to broadband, different national pricing structures and the online services offered (an aspect which has a bearing on how interesting and useful the internet is felt to be in the respective countries). Some of the factors affecting the extent of take-up are linked to potential demand, which in turn influence by income levels and ICT skills. In addition, without claiming that these are predominant factors, there will also be what might broadly be termed 'cultural' influences upon adoption and use. These have been initially charted in chapter one, but different facets are explored in chapters six and eight (see also Gilligan & Heinzmann 2005a, b).

One influence that may certainly have contributed to perceptions, attitudes and evaluations of the online world (which may in turn have had a bearing on adoption and use) is media coverage of the internet. To illustrate this, the *EU Kids Online* project was interested in perceptions of the risks facing children online and how parents act based on those perceptions, such as setting rules about what children can do on the internet – which in turn affects children's use (Hasebrink, et al., 2009). In principle, media representations could have a bearing on people's perceptions relating to other areas of the internet, such as eCommerce, eGovernment, online entertainment and leisure opportunities or, indeed, whether the internet is generally perceived as being a risky place to operate, not just for children.

---

[33] Although the authors collaborated the sections were predominantly written as follows: the introduction by Bergström, the methodology by Pascali, the methodological limitations by Gyenes, the results of the content analysis by Pinter and the conclusions by Haddon.

77

	UK	Italy	Hungary	Sweden
Internet access by households (Eurostat, 2008)	71%	42%	48%	84%
Broadband access by households (Eurostat, 2008)	62%	31%	42%	71%
Internet users (per 100 inhabitants, ITU 2007)	66%	54%	35%	77%
Reading online news, newspapers or magazines (all individuals, Eurostat, 2008)	37%	17%	33%	45%
Ordering goods or services for private purposes (all individuals, Eurostat, 2008)	49%	7%	8%	38%
Internet banking for private purposes (all individuals, Eurostat, 2008)	38%	13%	13%	65%
Reason for not having internet access at home: don't need internet ("Because not useful, not interesting etc.") (households without internet, Eurostat, Lööf, 2008)	30%	n.a.	49%	42%
Reason for not having internet access at home: lack of skills (households without internet, Eurostat, Lööf, 2008)	13%	n.a.	28%	40%

Table 1: National differences in experiences of the internet.

Media coverage may contribute to popular thinking about particular issues, such as debates about copyright in relation to downloadable material. Or sometimes the media may simply influence the symbolic impressions associated with a technology – its connotations, as explored in a previous COST project looking, cross-nationally, at social representations (Contarello, et al., 2008). In other words, there are a whole range of reasons for examining and levels upon which one can analyse such media coverage – asking how much is common across countries or whether there are national differences.

Media Studies research has developed a number of frameworks for examining such media processes. For example, one strand has focused on media moral panics, often specifically about children's relation to each new information and communication technology that appears – the internet included (Drotner, 1992; Boëthius, 1995; Critcher, 2008). Another strand of media analysis is that of 'agenda setting' whereby the media set the topics deemed to be of more public interest (McCombs & Shaw, 1972; Dearing & Rogers, 1996). Cultivation theory has looked at how media coverage can lead people to over-estimate the incidence of crime (Gerbner & Gross, 1976; Gerbner, et al., 1986). Here the emphasis is not on (the history of) particular media stories, as in the moral panic approach, but rather the routine processes of media coverage – in the example above, the routine coverage of crime that has a 'drip effect' of, over time, creating a (misleading) sense of the prevalence of the risk of crime.

In their various ways, all these approaches address the question of how the media frame reality (and potentially people's perceptions). In the research reported in this chapter the aim is certainly to address this issue of framing reality, and, in this chapter, this is achieved through a basic quantitative content analysis of media coverage, charting any convergences and variation across different national media.

As regards the specifically cross-cultural element, while there is an emerging literature on cross-cultural analysis, (Kohn, 1998; Livingstone, 2003) there is little on cross-national media differences. The work of Hallin and Mancini (2004) is probably the best known one for comparing whole media systems within Europe (e.g. the development of media markets, the degree of state interventionism, the extent of journalistic professionalism). But what is novel about the research project reported in this chapter is that it specifically examines variations in media content relating to a particular field: the internet.[34]

Methodology

The choice of countries studied – the UK, Italy, Hungary and Sweden – reflected the backgrounds of the participants in this study. Nonetheless, they represent not only geographical diversity in Europe but also the four media systems analysed by Hallin & Mancini (2004), and by Jakubowicz (2007). This means that the chosen countries represent different media markets, political systems, media systems and journalistic professionalism (as stated by Hallin & Mancini), all of which can be important factors for the actual output in news media.

---

34 Here there is a synergy with the *EU Kids Online* project, which has a strand focused on press coverage of children and the internet (Haddon & Stald, 2008).

The decision was made to focus on newspapers, since there were relevant databases of these in some participating countries or else the articles were relatively easy to collect for a set period (e.g. compared to monitoring TV coverage). The pilot study suggested that if each team collected articles from two newspapers for three months this would generate enough material for analysis. More than two newspapers would generate more material than could be handled within the time available.[35]

The initial pilot study was conducted in the UK, which also served as a first pilot in the *EU Kids Online* project mentioned above. This allowed the team to test and add to the coding system that was being developed. It also provided a sense of what material existed, provided some indication of the time periods that would be required to obtain different sized samples of articles, showed differences in results between examining the original paper copy of newspapers versus electronic copy held on databases and indicated the results of using different search words in those databases. Importantly, the initial pilot also revealed how much research effort, mainly in terms of time, would be required when following different strategies. In the Italy, Hungary and Sweden group participants conducted further pilot analyses of the press in order to evaluate these choices and see what further issues emerged.

A sub-sample of material from the UK collection was then used to test intercoder reliability, since all participants spoke English (for intercoder reliability in general see Lombard, et al., 2002).[36] On the basis of these discussions about the pilot feedback and inter-code test the coding framework and search processes were finalised.

It was decided that one of the newspapers would be from the quality press (in the UK this is traditionally known as a 'broadsheet'): the *Independent* in the UK, *Népszabadság* in Hungary, *Corriere della Sera* in Italy, *Dagens Nyheter* in Sweden. One possibility was to then collect material from one of the popular

---

35 To illustrate the trade-offs, in *EU Kids Online* the topic of children and the internet was much smaller, and hence more newspapers were examined in a fixed time period to generate a sufficient sample.
36 If we look at pairs of countries (e.g. UK vs. Italy), intercoder reliability ranged from 61%-85% for the centrality question (i.e. 'Is the internet the focus of the article?'). As regards the question of which section the article appeared in intercoder reliability ranged from 53% to 100%, but all the low scores turned out to be the same confusion of one particular pair of categories. In general there was considerable agreement. On the question of the area of life being examined intercoder reliability in the test ranged from 69%-85% - once again, in general there was a good deal of agreement. As regards the origin of the article, intercoder reliability in the test ranged from 38%-77%, but some of the low scores related to one particular process – while some classified an article as originating in one of the categories offered, others classified it as 'Other' if it had some extra nuance. Intercoder reliability for identifying the spokesperson ranged from 77%-85%. Some subsequent discussion to clarify coding was held.

press (or 'tabloids') in each country. But there was a question of what could be considered a tabloid in Italy – this is a first problem for comparative analysis, because in Italy this type of newspaper does not exist. However, since most countries seemed to have the free Metro newspaper, it was decided that this would be the second newspaper. Including the Metro added another dimension to the study, pointing to differences in content between the paid and the free press (Wadbring, 2007).

The time period chosen, based on the experience of the pilot study, was three months. The first collection was February, March and April, 2007.[37] A subsequent collection of articles took place in February, March and April 2008, both to explore whether there was much change on a year to year basis (or between different samples) and to have data that were nearer to the completion of the project. Only the first year's material is reported here.

In order to be comprehensive, a broad range of questions was asked, including the size of the article and whether there were illustrations (See Table 9 for the full list of questions[38]).

Methodological limitations

In the first year, the participants analysed and created a database of a total of 1457 articles. The individual totals are not reported since for Hungary, Italy and Sweden an electronic searchable version of the Metro was available online but this was not the case for the UK, whereas the reverse was true for the quality papers – the UK could use the press-database Lexis-Nexis to search for these, but in the other countries the quality newspaper was read in the printed form. This difference in search procedure produced some non-comparable totals, and some analyses were not possible (e.g. the UK database did not reveal if there were an illustration or not). But it seemed not to affect the other figures – for example, it was usually not the case that the UK was consistently different from the other three countries.

After collecting the data the decision was made to collapse together some categories listed in Table 9. For example, in the 'origins' of the story, the options, academic research, market research and other institutional research were initially separate as indicated in Table 9 but the percentages for each were so small that they were combined as 'research'.

Despite the precautions of conducting a pilot study, intercoder reliability testing and much discussion, it became clear when examining the findings that

---

37 Since this was decided at a meeting in early February, it was sometimes not practical to find paper copies for the earlier dates. In which case, some articles were collected in the first few days in May to compensate – e.g. in the UK.
38 A fuller analysis of all the variables considered can be found in Pinter, et al. (2009).

some options had been analysed slightly differently – e.g. in the 'area of life', 'culture' and 'technological developments' had clearly been open to broader and narrower interpretations depending on the analysts. Under 'origins' some analyses were based on a press release being inferred whereas others coded it only when a press release was explicitly mentioned. When this led to different statistics, it became clear that this was product of the classification process rather than a difference between national newspapers.

Finally, one last caveat is that each country chose one quality newspaper, and it is possible that in some respect choosing different newspapers would have had some effect on the statistics. That said, and partly based on a general familiarity with the various national media, the team felt that the general trends revealed in the figures reflected real similarities and differences in national media coverage.

Results of the content analysis

It is possible to conduct several types of analysis related to form and content both from a paid/free newspaper perspective as well as from a country comparison perspective. This chapter focuses on findings comparing the internet and broadband focus, and then continues by highlighting specific country differences. This means that single items of the data are picked to illustrate these differences and the presentation of our findings will not give the whole picture of the internet and broadband coverage in newspapers in the four countries.

As regards areas of commonality across countries, some of the findings confirmed the original expectations. Technology developments are not the most newsworthy aspects reported as the internet has become integrated into everyday life. Nowadays the online world is discussed over a range of other newspaper categories, the two most popular being security and crime (18% of all articles) and entertainment (17% of all articles).

The figures in the first line of Table 2 underline the way in which the internet is increasingly taken for granted: a third of the articles collected (ranging from 32% in the UK to 42% in Italy) refer to but were not primarily about the internet. Hence many readers now encounter references to the internet in the background or in passing.

In the second line of Table 2 it is clear that despite the fact that technology companies and policy makers currently have a strong interest in 'broadband', in practice this term only occurred in 6% of all articles.

	UK	Italy	Hungary	Sweden	Average
The internet is not the main focus	32%	42%	33%	36%	36%
Broadband is explicitly discussed	11%	5%	6%	4%	6%

Table 2: Percentage of articles where the focus is not the internet and where broadband is explicitly discussed.

It was also common across countries that references to broadband occurred mainly in discussions of the internet's technical infrastructure (56% of all articles) and the developing market for broadband (16%). In other words, broadband is framed as technological development and not as a social phenomenon – comment on the social consequences of the internet occurred in other sections of newspapers.

As regards commonalities, many of the options for the questions listed above attracted low percentages across the countries. Lastly, some media processes seem to be similar across countries. For example, academics/researchers appear as the spokespeople (experts) in articles ranging from 6% of articles in Hungary to 8% in the UK.

	UK	Italy	Hungary	Sweden	Average
National/ transnational law, regulation, statements	6%	5%	2%	2%	3%
Campaigns (lobbying, awareness raising)	7%	2%	0%	0%	2%

Table 3: Percentages of articles originating in laws, etc. and campaigns.

It is interesting to note that the national and European institutions seem unable to get their own views about the internet onto the media's agenda.

With the partial exception of the UK, shown in Table 3, they rarely become a news source for the media and they are not effective in using the media to communicate their own discourse concerning the internet, the broadband and innovation.

The same is true – again with the partial exception of the UK, in Table 3 – for NGO and civil society associations (which usually are the promoters of events connected with campaign, lobbying, awareness raising, etc.). Even if they have an agenda concerning internet and innovation and use the internet as a place for their initiatives they are rarely effective in creating news events and communicating actions that break the mainstream news-making routines.

Looking now at national differences, background factors such as the different level of internet adoption in each country seemed to make less difference to media coverage than variations in the media styles in the different national newspapers. This in turn, however, can be a consequence of the different media system models mentioned initially, which themselves include technological developments.

One notable exception is perhaps stories relating to e-commerce as an area of life: it is only covered in 4% of articles in Italy, which is 2-3 times less than in other countries in Table 4. Here media coverage probably reflects the fact that eCommerce has not been adopted so much in Italy, in large part because it is not trusted as a secure medium.

	UK	Italy	Hungary	Sweden	Average
eCommerce	12%	4%	8%	7%	7%

Table 4: Percentage of articles where 'eCommerce' was the area of life covered.

One observation concerns the organisation of newspapers: the same type of story can appear in one country's business news, in another's international news or in yet another country's technology sections. This potentially frames the subject matter slightly differently by virtue of where it is located (and thinking about the process of reading, some readers might not come across certain articles if they do not tend to read those sections).

For example, in the Swedish Metro, the proportion of entertainment items relative to the other countries is boosted by the regular 'website of the day' tip – but this is found in the financial section, even though it is not really 'financial' news! This single fact meant that the 'economy' section was relatively high in Sweden compared to the other countries. Meanwhile, in the Hungarian quality paper every fourth article was to be found in the technology section, which suggests that internet is still perceived as being a technological innovation rather than as an economic or socio-cultural phenomenon. In fact, in some of the other countries' press there was no dedicated section for technology. Conversely, in UK, 25% of all articles appeared in the section for company news - whereas the Hungarian Metro, for example, has no such section, so events like company takeovers tend to be reported in technology section in this newspaper, probably because it has limited impact economically on the everyday life of Hungarians.

In Italy, Sweden and the UK, the percentage of articles originating in the legal field is fairly similar (see Table 5), probably reflecting media routines where the police and the courts among the most traditional and common media sources.[39] But we see a noticeable difference in the case of Hungary, perhaps reflecting a lower incidence of crime reporting in general (and what coverage there is tends to be in terms of foreign news about crime and the internet).[40]

	UK	Italy	Hungary	Sweden	Average
Court cases, police actions or crimes	12%	11%	3%	10%	8%

Table 5: Percentage of articles where the origin was court cases, police actions of crimes.

Even where the percentage of articles originating from court cases, police action and crime reporting is similar, the framing varies in different countries. The clearest example is Italy, where the internet is more often framed as a "social problem" (Italy has the highest rate of articles characterised in this way with 11%, and Sweden the lowest with 1%), and where the newspapers tend to report the online world using highly emotive terms. Table 6 shows that in Italy a much higher percentage of members of the public and victims find a voice in the media than in other countries, but conversely there are also more institutional voices talking about the internet (even when it is related to crime, police action, etc.). This of course is relevant to the construction of a more widespread apprehension about the online world in the Italian press.

	UK	Italy	Hungary	Sweden	Average
Member of the public, victim	13%	25%	20%	14%	18%
Non-commercial Institutions	8%	12%	7%	7%	9%

Table 6: Percentage of spokespeople quoted in new stories.

---

39 In the *EU Kids Online* study relating to press coverage of children, this source was even higher and dominated coverage (Haddon & Stald, 2008).
40 Generally, not only in the case of legal/crime related internet news, there is much more foreign internet news reported in Hungary than national internet news.

What is perhaps striking about the UK newspapers is, in various guises, the presence of business in articles about the internet. There is more business internet news than in the other countries and we more frequently hear the voices of company spokespeople.

	UK	Italy	Hungary	Sweden	Average
Brand mentioned	30%	13%	12%	12%	15%
Spokesperson: Company[41]	48%	18%	24%	42%	33%
Origins: Company reports, company statements, profit warnings	32%	1%	1%	1%	6%

Table 7: Percentage of stories mentioning brands, having company spokespeople and originating in company reports.

Nearly half of the articles mention at least one brand in the UK, and many different companies mentioned one. In other countries only approximately 15% of the articles contain brands. Finally, a much higher proportion of the stories originate from companies reports in the UK. Most probably it can be explained by the fact that London has a strong stock exchange and vivid business life; hence the presence of companies in the media has had a long tradition.

Another difference is related to the higher level of 'political' presence in the news in Italy and Hungary. In the Italian case this is mainly connected to the habit by journalists of reporting politicians' points of view on almost every kind of issue, which in turn relates to the high degree of mediatisation of Italian politics. In the case of Hungary the reason is the ongoing debate on the role of Hungarian Government in building the information society and propagating internet use because of the current relatively low penetration rate.

	UK	Italy	Hungary	Sweden	Average
Politicians, government	6%	17%	19%	8%	13%

Table 8: Percentage of politicians and government spokespeople in articles.

---

41 Internet industry, media industry and other companies.

In contrast, in Sweden there were no 'hot' political issues about broadband and the internet by the time of the data collection. Five to ten years ago there probably would have been, with focus on how to overcome the digital divide, how to organise broadband infrastructural development, but nowadays the country no longer faces these kinds of challenges.

Conclusion

This study has focused on the meaning of the internet, since what it symbolises (at various levels) has a bearing upon perceptions of how much the internet is leading to changes in the way we live our lives, to changes in the economy, the political system and cultural life more generally – and whether this is for better or worse. What we understand by the internet and what is happening on it, or what we can do with it, can have a bearing upon people's decisions to use it, their awareness and choice of what to do online, what they let their children do, etc. In other words, the symbolism of technology is important, important for understanding usage, but important for reasons beyond that.

People derive their understanding of the internet from various sources, but one key source of representations of life online comes from the media, increasingly so according to some of the theorists discussed. Hence, this chapter has examined the levels on which the media may provide audiences with messages and whether and how these vary cross-nationally. One of the key interests of this book is the cultural factors at work, especially affecting adoption and use. Here, media representations can be viewed as being one such cultural influence.

Although, in principle, the interest is in media more generally, the focus has been on the press both for practical reasons and because reading figures generally indicate that this medium retains an important role in the life of many Europeans. One limitation is the decision to cover just two newspapers per country. However, the parallel *EU Kids Online* study drew on a sample from more newspapers per country and also found find data showing cross-national differences. Hence, even with this limited sample size of the project report here, when coupled with our background knowledge of the national media of these four countries we believe that the findings do reveal something about country differences and certainly about different media logics in the press.

To start with the commonalities across the different national newspapers the evolving internet still remains somewhat special in the sense that stories are often reported because they involve the internet. Of course, that can be true of other media, such as stories about developments in television, but at this point at least the majority of coverage still places it in the foreground. That said, there is also a process of routinisation since the internet is now only in the background in many stories as it has become more taken for granted. Meanwhile, it is the internet in general that is still special, as indicated by the number of stories about internet firms, websites and emails, although its continuing evolution through developments such as web 2.0 applications may help to keep it newsworthy. In comparison, broadband is only explicitly discussed in a small proportion of cases, and mainly in relation to technological and market developments rather than social ones.

Aside from these commonalities, this chapter has demonstrated the numerous ways in which, and levels of which, some country variation exists. In that sense has contributed to cross-cultural analysis of the media more generally, complementing the comparative analysis of media systems, as well as more substantively adding to our understanding of representations of the internet.

Turning to the detail, the variation examined included the coverage of particular topics (e.g. eCommerce), the area of life from which stories are drawn (e.g. crime), the origins of articles, and the spokespeople whose voice is heard in the different media.[42]

Of course, some of the variation in coverage may reflect aspects of the society more generally, for instance certain values, in which case there is an argument that the media are reflecting the wider social context.[43] However, we saw how national internet diffusion rates did not seem to shape the level of coverage overall and we strongly suspect that the different media logics discussed earlier count for much of the variation. Thus point is made all the more clearly when formatting differences between national newspaper are such that certain sections do not even exist in some of the press covered. Of course, a quantitative content analysis alone cannot explain all the reasons for the cross-national variation identified, but we have tried to provide some plausible suggestions from wider background knowledge of the national media covered. In other words, the chapter has attempted to contextualise, indeed account for, some of this variation through observations about the, especially cultural, processes at work in the respective countries.

To acknowledge another limitation, a study that looks at coverage alone cannot in itself prove that this affects readers' understandings of the online world. But media theories suggest that the way in which the press frames

---

[42] Although it is not covered here, there was also variation in how the articles visually appear, as reflected in their number, size and whether they are illustrated (see Pinter, et al., 2009). This can - even if only at an unconscious level - affect our impressions of the significance of the online. It can involve variation in the framing of articles by virtue of the sections in which they are located within newspapers.

[43] An example of this from *EU kids Online* was the question of whether the lower concern about pornography in Norway reflected a particular view of the child and sexuality more prevalent in the Nordic countries.

coverage of phenomena does have important implications for how it is perceived in the wider society.

Question area	Rationale
Which part of the internet is discussed? 1. Internet in general 2. Websites and world wide web, domain names 3. Internet and computer infrastructure: standards, software, hardware, wires, types of connection 4. Internet activities, services and economy: e-banking, e-commerce, online shopping, internet advertising, marketing, searching and search engines, e-work 5. Internet business: company take-overs, profitability, price of internet, size of markets etc. 6. Education and research: e-learning, blended learning, statistics, online survey, etc. 7. Politics, democracy and administration: e-voting, e-petition, e-government (downloading forms), e-democracy, regulation, censorship, etc. 8. Entertainment and media: online gaming, downloading, gambling, virtual worlds, sexuality, digital TV, new media, online newspapers, radio etc. 9. Communication: e-mailing, IM, chat, VoIP, forum, videoconference, etc. 10. Web 2.0: blogging, citizen journalism, podcasting, Wikipedia, file-sharing, video-sharing, photo-sharing, social networking sites, etc. 11. Security and crime: security, privacy, virus, spam, adwares, cyberbullying, phishing, hacking-cracking, sexual predators, paedophilia, etc.	The aim was to see if any aspect of the internet currently had more visibility in the media.

Is Broadband explicitly discussed? 1. Yes 2. No	It was important to check this given that policy makers often use this term.
Is the internet the focus of the article? 1. Yes, internet is the centre of the article 2. No, the internet is only discussed in passing, it is not the focus	To what extent do people encounter the internet as a background feature when reading about other things?
Which section contains the article? 1. Local news section 2. National news section 3. International news section 4. Politics section 5. Lifestyle section 6. Humour section, anecdotes 7. Job section 8. Economy section 9. Technology section 10. Money/saving section, Product comparison section 11. Travel section 12. Personal advice section 13. Frontpage 14. Letters 15. Competitions (reported on or run by the paper) 16. Interpersonal, dating section 17. Entertainment section 18. Education 19. Sport 20. Radio/TV 21. Editorial/debates/opinion	The section in which an article is located can frame the internet story - e.g. whether it is within a product comparison section or humorous one.
What area of life does it relate to? 1. Technology developments 2. Legal, crime/ police/courts 3. Hacking 4. Citizen's rights 5. Work 6. Education 7. Entertainment 8. Sport 9. Politics 10. Medical	For example, do we mainly encounter internet stories relating to education or to business?

11. Interpersonal/sexual relations 12. Banking 13. e-commerce, online shopping 14. Security industry 15. Media 16. Travel 17. Product comparisons, shopping 18. 'Human Interest' story 19. Social problems 20. Environment 21. Personal reflection 22. Betting, gaming, gambling 23. Culture	
Is there any special origin of the article (a source that generated it), such as: 1. Academic research 2. Market research 3. Institutional (official) research 4. New national or transnational law, regulation, statements 5. Academic events (e.g. conference) 6. Market events (fair, trade show) 7. Company reports, company statements, profit warnings, etc. 8. Press conference, press release 9. Campaign (lobbying, awareness raising) 10. Court case, police action and crime reporting 11. Other	What type of events generate articles – for example, how important is research in generating media coverage?
The views of what agency/ spokesperson, if any, is reported: 1. Internet industry 2. Politicians, government 3. Media industry (apart from internet) 4. Legal representatives, police 5. NGOs, charities 6. Researchers, academics 7. Medical representatives 8. Trade associations 9. Celebrities 10. Member of the public, victim	Traditional media analysis has often been interested in the question of whose voice is heard in the media, who is quoted, who is sought out as a spokesperson.

| 11. Consumer groups
12. Other companies (not media, internet)
13. Institutions (non-commercial)
14. Education | |

Table 9: Selection from the content analysis coding system

Bibliography

Boëthius, U., 1995. Youth, the media and moral panics. In: J. Fornäs & G. Bolin, eds. *Youth culture in late modernity.* London: Sage, pp. 39–57.

Contarello, A., et al., 2008. ICTs and the human body: an empirical study in five countries. In: E. Loos L. Haddon & E. Mante-Meijer, eds. *The social dynamics of information and communication technology.* Aldershot: Ashgate, pp. 25-38.

Critcher, C., 2008. Making waves: historical aspects of public debates about children and mass media. In: K. Drotner & S. Livingstone, eds. *The international handbook of children, media and culture.* London: Sage, pp. 91-104.

Dearing, J. W. & Rogers, E., 1996. *Agenda-setting.* Thousand Oaks: Sage.

Drotner, K., 1992. Modernity and media panics. In: M. Skovmand & K. Schrøder, eds. *Media cultures: reappraising transnational media.* New York and London: Routledge, pp. 42-62.

Eurostat, 2008. *Internet access and use in the EU27 in 2008.* [Online] STAT/08/169. Available at: http://europa.eu/rapid/pressReleasesAction.do?Reference=STAT/08/169 [Accessed 18 December 2009].

Gerbner, G. Gross L. Morgan, M. & Signorielli, N., 1986. Living with television: the dynamics of the cultivation process. In: J. Bryant & D. Zillman, eds. *Perspectives of media effects.* Hillsdale, N.J.: Lawrence Erlbaum Associates, pp. 17-40.

Gerbner, G. & Gross, L., 1976. Living with television: the violence profile. *Journal of Communication,* 26 (2), pp. 173-99.

Gilligan, R. & Heinzmann, P., 2005a. Exploring how cultural factors could potentially influence ICT use: an analysis of Swiss TV and radio use. In: L. Haddon, ed. *International collaborative research: cross-cultural differences and cultures of research.* Brussels: COST, pp. 87-96.

Gilligan, R., & Heinzmann, P., 2005b. Exploring how cultural factors could potentially influence ICT use: an analysis of European SMS and MMS use. In: L. Haddon, ed. *International collaborative research: cross-cultural differences and cultures of research.* Brussels: COST, pp. 97-116.

Haddon, L. & Stald, G., 2009. *A cross-national European analysis of press coverage of children and the internet*. [Online] Available at: http://www.lse.ac.uk/collections/EUKidsOnline/Reports/MediaReport.pdf [Accessed 18 December 2009].

Hallin, C., & Mancini, P., 2004. *Comparing media systems: three models of media and politics*. Cambridge: Cambridge University Press.

Hasebrink, U. Livingstone, S. & Haddon, L., 2008. *Comparing children's online opportunities and risks across Europe: cross-national comparisons for EU Kids Online*. London: London School of Economics, EU Kids Online.

ITU, 2007. *ITU Statistics from 2007*. [Online] (Updated 18 December 2009) Available at: http://www.itu.int/ITU-D/icteye/Indicators/Indicators.aspx# [Accessed 18 December 2009].

Jakubowicz, K., 2007. The Eastern European/post-communist media model countries. In: G. Terzis, ed. *European media governance: national and regional dimensions*. Bristol: Intellect Books.

Kohn, M., ed., 1998. *Cross-national research in sociology*. Newbury Park: Sage.

Livingstone, S., 2003. On the challenges of cross-national comparative media research. *European Journal of Communication*, 18 (4), pp. 477-500.

Lombard, M. Snyder-Duch, J. & Campanella Bracken, C., 2002. Content analysis in mass communication. Assessment and reporting on intercoder reliability. *Human Communications Research*, 28 (4), pp. 587-604.

Lööf, A., 2008. *Internet usage in 2008 – households and individuals*. Eurostat Data in Focus 46, 2008. [Online] Available at: http://epp.eurostat.ec.europa.eu/cache/ITY_OFFPUB/KS-QA-08-046/EN/KS-QA-08-046-EN.PDF) [Accessed 18 December 2009].

McCombs, M. & Shaw, D., 1972. The agenda setting functions of the mass media. *Public Opinion Quarterly*, 36 (1), pp. 176-87.

Pinter, R. Bergström, A. Gyenes, F. Haddon, L. & Pasquali, F., 2009. *Give us our daily broadband – a comparative press discourses project in 4 countries*. A report for COST298.

Wadbring, I., 2007. The role of free dailies in a segregated society. *Nordicom Review*, 28, pp. 135-48.

Leslie Haddon and Peter Heinzmann

Chapter five. Implications of the variation in broadband speeds over time

Introduction

Many factors affect people's interest in acquiring and using new information and communication technologies (ICTs) and specifically new applications. Such factors include, for example, economic costs, network effects in the case of communications (i.e. how many other people use an ICT), the symbolic meanings associated with ICTs whereby it becomes fashionable to have some technologies, etc. This report, though, aims to reflect more on the features of technology itself, specifically its capability to support certain functionalities and, indeed, to do so in a way that makes them if not attractive, then at least acceptable. In particular it looks, historically, at various changes related to the experience of the speed of broadband[44] and how this has influenced users' experience of different applications and hence their different uses of the internet.

Some writers examining the topic of the digital divide have made the point that one should not just look at whether different people are internet users or not but also consider how the online experience is affected by factors such as their technical skills, their awareness of online possibilities, the support of their social networks – and the very technology through which they experience the online world (Bakardjieva, 2001; Haddon, 2004). It makes a difference, for example, if one is using older or newer equipment, with differences in processing power or one subscribes to ISPs with faster or slower connection speeds. Looking at the technologies and connections people use gives a more nuanced picture of whether some have more advantages over others, and how the affordances of people's technologies affect the quality of experience they can have of the online world.

Applying this approach beyond the digital divide debates, one can simply say that certain usages that have now emerged would have been impossible, or less

---

44 Since technologies are always changing, what counts as 'broadband' also continues to evolve - indeed, most reports using the word 'broadband' do not provide a precise definition of the term. Today, the term broadband typically includes widely used internet connections 500 times faster than early internet dial-up, i.e. 'narrowband', technologies 10 years ago. However, the term broadband does not refer to either a certain speed or a specific service. In this chapter we understand by 'broadband connections' internet connections with a download speed higher than 128kbit/s, with flat-rate (non-volume based) tariff charges for fixed landlines and volume based charges for mobile phones.

practical, or certainly less attractive, when connection speeds were slower. However, this is not just about looking at the changeover from narrowband[45] to broadband. From their introduction, although the offerings of different ISPs may have all been called broadband, the actual speeds of connections varied.

Let us consider some examples. Even with narrowband one could send images, but slower upload speeds meant that this involved spending some time sending those images to others or posting those images online (which could be both tedious as well blocking telephone lines for voice calls in many homes). Early broadband at around 600 kbit/s download and 100 kbit/s upload speed certainly made sending and receiving single images much quicker, but more speed was needed to make the practice of sending (and receiving) many images viable. For example, in interviews in Korea in 2005 some students were posting many photos of friends that they had taken with their camera phones on, say, a group excursion, where a 3 day trip might have produced something in the order of 200 pictures (Haddon and Kim, 2007).

Or take websites. While they have been around for many years, earlier versions contained mainly and almost exclusively text, compared to the websites of today, which have many different objects and sometimes extensive audio-visual material. At what point in time (and for whom, if we take a cross-national perspective) did the changing speed of broadband support practices such as posting many images and managing complex websites?

Meanwhile, one study of Italian youth discussed how they commonly share online music, TV and film material (Mascheroni, et al., 2008). Again, this has become more feasible as speeds have evolved. But what was the tipping point to allow this?

In calculating how people have experienced connection speeds and hence download and upload times, we note that the headline speeds - i.e. the topspeeds claimed by the ISPs who supplied those connections - provide just a starting point. Others have also demonstrated (e.g. Ofcom 2008, 2009) that the reality of speeds experienced by users for various reasons does not always match the potential. Hence, empirically, we now examine historical data about the actual average speeds experienced at different times, and the pattern by which those speeds evolved, since this had a bearing upon what applications (and the hence what practices of users) could be supported or could become acceptable.

Lastly, averages conceal the variation in people's experiences. For example, the length of the connection line and congestion at certain peak times both influence speeds. Hence we have to consider what effect these have on transmission times in order to appreciate the range of experiences within countries at different points in time. Finally, while we can offer examples of single countries, trans-

---

45 In this chapter 'narrowband' is understood to mean the dial-up connection to the internet with speeds of less than 128kbit/s, being charged on a time or volume basis.

mission rates varied significantly in different European countries. So how much of a gap is there between the speed experienced by people in different countries and how has this gap changed over time? Has it become larger or smaller?

Thus the rest of the chapter proceeds through the following stages, providing:
– a first impression of the time that would be taken to upload and download files given different theoretical connection speeds
– an explanation of the methodology behind the next stages of the chapter.
– an illustration of how file sizes have grown over time for different applications and changes in streaming speeds for different applications.
– an historical overview of how long it took to transmit typical files in different years.
– a view of the difference between headline speeds (over the years) and the actual speeds people experienced, with implications for transmission times.
– a discussion of variations within countries and why these occur, with what implications.
– an overview of the differences in connections between countries, and how those differences have changed over time.

Methodological issues

Streaming and 'bursty' traffic

There are two basic types of information transfer in telecommunication and on the internet leading to streaming and 'bursty'[46] traffic.

In the case of *streaming traffic* the receiver already consumes the first parts of information while other parts are still being sent by the sender. Streaming traffic involves the regular delivery of the pieces of information. For example the digitised audio signals that are common in conventional telephony require the regular delivery of one byte (8 bits) of information every 125 microseconds. This leads to a constant average data rate of 64 kbit/s (for the entire duration of the transmission). Another example is streaming audio or video information on the internet, as for example when people are listening to internet radio programs or when they are watching YouTube videos leading to an average data rate of 128 kbit/s and 900 kbit/s respectively. Here, compression is used and therefore the instant data rate may vary. But again, we have a rather constant data rate during the whole transmission.

Bursty traffic is irregular; it comes in bursts. Sometimes there is a need for communication at typically high data rate, sometimes not. Hence the average data rate during a so called 'session' – that is, an extended period of communi-

---

46 Bursty traffic refers to an intermittent pattern of data transmission.

cation between two parties – is much lower than the peak data rate. In such bursty traffic situations the communication channel is typically shared by many users. While one user is not requesting information the channel is free for other users. This type of traffic is typical in computer communication networks as for example when people access web servers, in computer games, or when people do access remote computers.

Response time and data rate

In the case of bursty traffic the 'response time' a user experiences when requesting a piece of information is the most important technical performance indicator. The response time requirements depend on the 'use case' i.e. the type of application.

In the case of streaming traffic, however, it is the available data rate of the communication channel that is the most important technical performance indicator. If the streaming information belongs to a one way communication channel such as a radio broadcasting signal, then the response time – that is, the time it takes from the mouse click requesting the programme to the moment in time when the sound can be heard at the receiver side – is not very demanding (e.g. it can be up to a value measured in seconds).

But with streaming (e.g. audio signals belonging to a telephone call) the so called 'round trip time' must be less than 350 milliseconds in order to have a fluent conversation. When it is longer, the two conversation partners feel uncomfortable because of the delayed reactions of the other side. We are familiar with such longer round trip times from intercontinental telephone calls that are routed via satellite, where the time from sending out a phrase until the reaction from the conversation partner arrives is typically around half a second.

Depending on the use case, i.e. the application, there are three classes of bursty traffic to be distinguished: batch processing, human interactive and machine interactive.

In the case of 'batch processing', response times of tens of seconds to hours may be perfectly acceptable. When requesting the transmission of a 'batch' of information, for example sending an e-mail or making a backup of files, there is usually no need for an instant response. However, in the so called 'human interactive' applications, people want an average response time of clearly less than one second. Moreover, this response time should be fairly stable - it should have little variation. Human interactive applications include, for example, web browsing, database queries or remote access to computers. Finally, in the 'machine interactive" applications computers interact with each other.

The next step involves looking at what shapes response times. One factor that determines the response time is the available transmission speed (the data rate) between the source of the information (e.g. a Web server) and the destination or 'sink' for the information (e.g. a user's client program). The other important factor is the volume of the piece of information requested.

Object Type	Data Volume	Data Rate				
		33.1 kbit/s	600 kbit/s	2 Mbit/s	5 Mbit/s	20 Mbit/s
Web page	500 kByte (4,000 kbit)	2 min	7 sec	2 sec	1 sec	0.2 sec
4 minute music track	4.6 MByte (38 Mkbit)	19 min	63 sec	19 sec	8 sec	2 sec
2 minute video clip (352x288px)	20 MByte (158 Mbit)	1.3 hours	4 min	1 min	32 sec	8 sec
143 minute DivX quality film (592x312px)	940 MByte (7,876 Mbit)	3 days	3.6 hours	66 min	26 min	7 min
143 minute DivX quality film (720x384px)	2 Gbyte (15,938 Mbit)	6 days	7.4 hours	2.2 hours	1 hours	13 min
143 minute HD quality film (1280x688px)	8 Gbyte (66,605 Mbit)	23 days	31 hours	9 hours	4 hours	56 min
Blue-ray disc (2 hours 1080p full HD quality film	50 Gbyte (419,430 Mbit)	147 days	8 days	2.4 days	1 days	6 hours

Table 1: Theoretical time taken to transmit different application objects[47].

In order to convey a first impression of what data volume and data rates can mean to our use of the internet, Table 1 shows the minimum response time when requesting typical 'application objects' (i.e. items we might want to send or receive online) on networks with different data rates. The application object data volume refers to data volume sizes used in 2009.

This is a straightforward table, showing that if we have files of different sizes on the left (presented in an approximate and more exact way[48]) and different data rates (that is, connection speeds) on the top, then this is how long it will

---

47 Similar tables are also available in other reports, as in the one from the British regulator Ofcom 2008, p.14 (Ofcom, 2008).
48 Note that sometimes a 'Byte' is abbreviated with a capital 'B' and a 'Bit' is abbreviated with a small "b". One Byte consists of 8 Bits.

take to transmit the files – either to download or upload them. It is clear that for some file sizes, as in the 250kB web page, one reaches the 'human interactive' stage very quickly, after 600kbit per second (kbit/s), where the user would probably find the speed acceptable. In contrast, for large file sizes, such as downloading DVD Quality Movies, the effect of faster speeds on download times at each stage in the evolution of broadband is more noticeable. In fact, at the very slow response times most people would not have bothered to download movies, video clips and probably even music. In those cases we are clearly in the 'batch processing' mode of traffic, or it might be even faster to physically transport the discs.

But these are all just typical 'example' values, even if they provide a rough guide. If one asks for the actual time it takes, for example, to get an actual video clip one must consider many more details. The size of the video clip file may vary considerably not just depending on the type of file but also on the year. The volume of a single web page is just one issue. But in 2009 most web pages consist of 10s to 100s of objects which may be fetched from multiple servers and therefore additional delay when accessing other servers may be the factor which mainly determines the overall response time. And we will demonstrate that all file types tend to get larger over the years. Finally, taking into account the actual speeds we will come up with figures and explanations which describe the users' actual experiences more realistically than the advertised speeds.

Next it is important to consider file sizes: these vary depending on the application and the file sizes for particular applications themselves change over time. In order to get some idea of current average file sizes as well as about historical evolution of file sizes, information about file sizes was assembled from a number of sources. Agrawal, et al (2007) have analysed metadata from over 10,000 file systems on Windows desktop computers at Microsoft Corporation over the time period 2000 to 2004. In addition, these data are supplemented by an analysis of the files on the file store of a small IT company with ten employees for the time period 1997 to 2009.

Lastly, one project within the COST298 action whose chapters formed the basis of this book had the task of measuring broadband performance. This involved installing software on "reference servers" in different countries so that internet users could select a web page on their country's reference server and check their connection speeds. The resulting information was also kept centrally at the cnlab in Switzerland, so that over time a considerable number of measurement results were assembled. For the case of Switzerland it is estimated that about 10% of the internet users have performed such measurements.

This self-selecting group of people who check their speeds is by no means a random sample of users. For example, people who check their speeds probably tend to be more enthusiastic about the technical aspects of the internet and/or are more advanced users. They may well be more likely to pay more for faster internet speeds. Alternatively, people who experience speed problems may run more tests than people who are satisfied with their connection speed.

The point of the performance test platform was to compare the connections of different ISPs and to analyse regional differences in certain countries. The objective of this chapter is to see how speeds change over time, so it does not matter so much if the people running tests are a little ahead of the rest of the population – it just gives a broad indication of the wider trend and highlights some issues that others would also experience.

In Switzerland the facility is used by about 2000 users per day carrying out around 10,000 tests per day. This is due to the media visibility of the project, which has not happened yet in other countries. Since there are more data about users in Switzerland and these have been collected over a longer period of time, the Swiss data have been used to demonstrate many of the processes at work, especially for the earlier years of broadband.

Where a comparative data is introduced this is also from the performance measures outlined above. Finally, in the discussion of variation in the experience of speeds within countries, UK data from the British regulator is used to supplement the main observations. We hope to further promote the performance test in various countries and hence to get much more users testing their connections in other countries than just Switzerland.

The time taken to transmit different file sizes

In the first section we take the first step of looking, empirically, at the current and historically changing file sizes and available data rates examining the implications for the time taken to download these files. At the moment we consider only 'headline speeds' – i.e. what is the time taken if we assume the maximum speed of connection cited for particular years. In the next section we will then go on to look beyond this to consider factors affecting the actual speed that users experience, and the implications for download times. Table 2 shows the current (2009) average file sizes for different activities and applications have been assembled from various sources. It also indicates the time needed for the transmission of the different file types using the current widely used 5Mbit/s internet connection.

Object, Application	Average files size 2009 (MB)	Transmission duration at 5Mbit/s speed
Full disc back-up	80,000	2,185 min
Video clip (.avi format)	942	26 min
Windows Vista. Service Pack 2 (SP2), May 2009	348	10 min
Video clip (.mov, .mp4, mpg format)	40	66 sec
Windows Media Player (wmp11)	25	40 sec
Vido clip (.wmv)	23	38 sec
Music files (.mp3 format)	5.6	9 sec
File Hosting Rapidshare/Mega upload normal	5.0	8 sec
Photos on a camera	2.0	3 sec
Photos on the nobile phone	0.8	1 sec
Web pages	0.5	1 sec

Table 2: File sizes for different objects / applications[49]

In Table 2 we can see that backing up an entire hard disk takes hours at 2009 speeds, whereas files that update software take minutes. Other transmissions need only seconds. Sometimes we have had to add several different possibilities, for example, in the case of video. This is because the resolution, the frame rate and the duration of videos clearly determines file size, and here we provide our best judgement as regards average file size. But the format also affects size; the file from a video clip of a given duration in the (uncompressed) .avi format is much larger than in the compressed .mpeg or .avi format. Music files are, understandably, smaller then video ones, and photos take up even less space, the file size between digital cameras and mobile phone photos being typically different because of the different resolution. File hosting refers to the files or pieces of files one can store on the internet instead of on one's PC, which is becoming an increasing popular way of sharing audio-visual material as opposed to exchanging peer-to-peer. Ultimately, Table 2 gives us a first idea of the time it takes to, usually download, different types of file, if we have the full speed of 5Mbits/s that the ISP claim to provide.

---

49 Statistics were taken in 2009 from files on the file servers of a small company.

The changing size of files and response times

While Table 2 showed the current average file sizes, these sizes have changed over the years – mostly they have increased. For example, the upper line of Figure 1 shows how home pages grew in size from June 2006 to December 2007 (17 months). This represents a growth of 34% between those dates, which means 24% a year – a growth rate of just under a quarter. This leads to larger files doubling in size in 3 years or growing tenfold over 10 years.

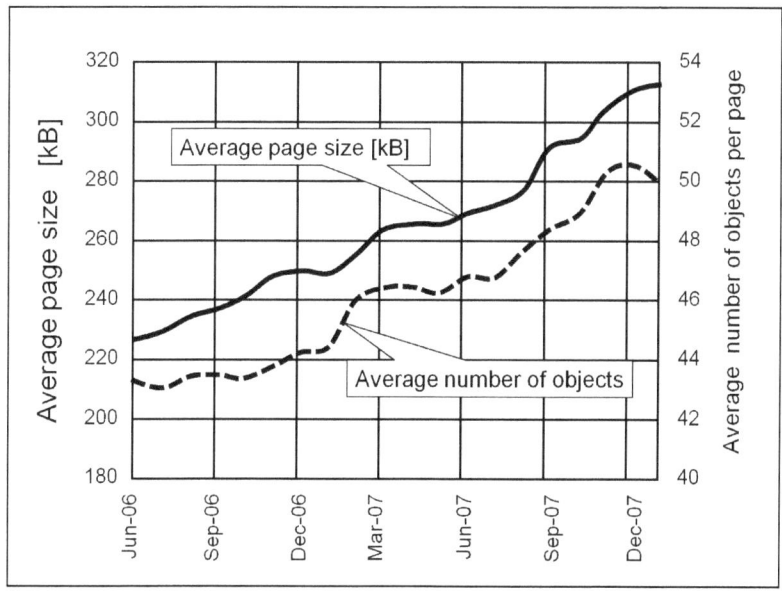

Figure 1: The changing size of home pages.

Taking an even longer term perspective, *web video clips* distributed over the internet have on average become longer (in terms of duration) over a decade. While the median web video clip duration was 45 seconds in 1997, it had grown to 120 seconds in 2005 and 193 seconds in 2007 (Growth of the Average Top 1000 Home Page, 2008). Hence the video clip has grown fourfold in duration over 10 years. But at the same time, the image resolution and the frame rate have also increased, leading to an even higher video clip file size growth over the last 10 years.

Figure 2 shows how the average office file sizes – Word (designated by 'doc') Excel ('xls'), PowerPoint ('ppt') and the exe-file sizes – grew over the last ten years[50]. For instance, PowerPoint files have grown about 20 times over the last 10 years, mainly due to the increasing use of images in PowerPoint charts[51].

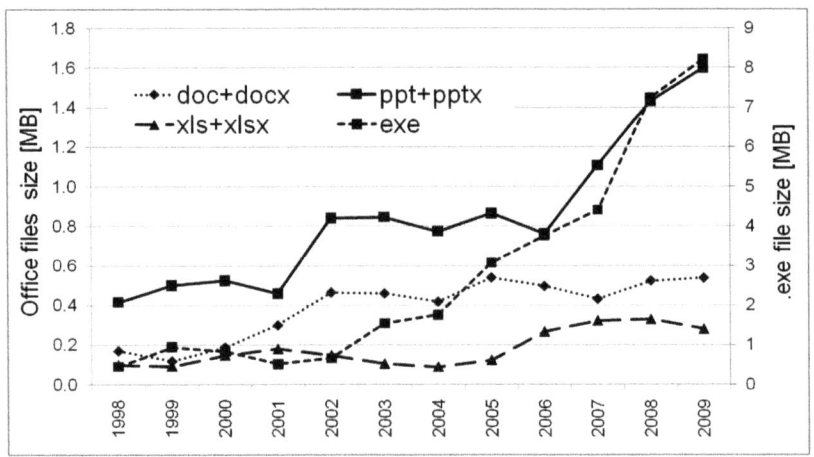

Figure 2: Average Office file size growth over time.

Modes of operation (user behaviour)

Obviously requesting a larger file will lead to larger response times than a smaller one. In the 'user interactive' mode people wait for the information they requested. When really working interactively e.g. remote computer access or surfing web pages users do need the information within sub-seconds. They may accept seconds for images or video clips appearing in web pages. But as soon as

---

50 These data were taken from the file server of a small company with 10 employees.
51 Haddon and Heinzmann (2009) also chart the growth of Adobe Acrobat Reader, Windows Media Player and Skype, the voice over IP package. If we look at Windows service pack and Windows patch file sizes theoretically downloading a 150MB service pack over the network in 2002 would have taken almost 5 hours at the 64kbit/s speeds broadly used at that time. No wonder that the software was in practice distributed on CDROMs. Today, the 350MB to 450MB service packs can be downloaded within about 15 minutes thanks to the current 5Mbit/s speeds. Thus, use of the frequent online software updates and patching that it is common today would simply not have been practical at the download data rates available five years ago.

things take longer then tens of seconds, e.g. when downloading documents or applications, people tend to switch to other tasks rather than waiting for the information. Hence, in effect, they change to batch processing mode.

Note that when information is delivered in streaming mode (e.g. video and audio) users get the beginning of the information before the whole file has been transferred; hence the response is faster than the time it would take to transmit the whole file.

To give an example of what this means in practice, at speeds of 5Mbit/sec a 2 minute video clip would take 32 seconds to download and a 143 minutes DivX quality movie would take 26 minutes to download. But playing the movie in streaming mode will show the first scenes within seconds. We will return to these examples as benchmarks in the later sections.

Symmetric and asymmetric communication

Traditionally most internet users are mainly consuming information, i.e. browsing web pages, downloading documents and programs, listening to audios and watching videos. Therefore ISP connection speeds are very asymmetric, having typically around 10 times smaller upload speeds than download speeds. For example, the upload speed for a line with 5Mbit/s is only 500kbit/s. Hence it would take about 3 seconds to download a single picture taken by a cameraphone but 32 seconds to upload it.

Around 2006 more and more internet users began producing and sharing information – writing information into forums or wikis, or uploading photos or videos to information sharing platforms like Flickr or YouTube. In this context the terms 'web 2.0' or 'read/write web' and 'prosumer'[52] were created. Until 2009 the so called 90-9-1 rule was almost certainly correct – i.e. the claim that in most online communities, 90% of users are lurkers who never contribute, 9% of users contribute a little, and 1% of users account for almost all the action. But the fame of information sharing platforms like YouTube, Flickr, Wikipedia or Twitter and the omnipresence of buzzwords like Blogging (e.g. Technorati), Social Bookmarking (e.g. Delicious), Social (e.g. Facebook, MySpace) or Business Networking (e.g. Xing) clearly point towards a growing amount of symmetric communication and hence more symmetric communication channel needs. Internet telephony (Skype) and video telephony applications provide further examples of such growing symmetric traffic.

---

52 The buzzword `Prosumer´ has multiple conflicting meanings. Here it is meant as the combination of information producer and consumer (as introduced by Alvin Toffler in 1980. In the business sector the prosumer is the market segment of professional consumers.

Historical evolution of transmission speed and file size

Clearly in the past, when transmission data rates were slower, it would have taken longer to download the current versions of the various files listed above. On the other hand, we were simply not sending such large files in the past.

Figure 1 and Figure 2 illustrate how, for different types of applications, the files were much smaller then. So what we can now ask is, given speeds at different time points but also the fact that there were different sized files at those time points, has the response time changed – are people able to download these items faster?

Year	Download data rate kbits/s	Web pages		Office files (Power-Point)		Photos	
1997	34	20 kB	5 sec	0.4 MB	2 min	0.8 MB	3 min
2001	64	60 kB	8 sec	0.5 MB	1 min	0.8 MB	2 min
2005	600	150 kB	2 sec	0.8 MB	11 sec	1.5 MB	20 sec
2009	5,000	500 kB	1 sec	1.6 MB	3 sec	3.6 MB	6 sec

Year	Download data rate kbits/s	Music song		Software (e.g. Adobe Reader)	
1997	34	1.0 MB	4 min	4.5 MB	18 min
2001	64	2.0 MB	4 min	8.4 MB	18 min
2005	600	3.0 MB	41 sec	16.3 MB	4 min
2009	5,000	7.0 MB	11 sec	26.6 MB	1 min

Table 3: Historical data showing file size and response times for downloads.

In Table 3 on *downloading* we can see that the trend across most files was that file sizes grew larger, but download speeds increased by an even greater factor so that the actual times for downloading all these files was reduced. Even in 1997 web pages with file sizes typical at that time could be downloaded fairly quickly and so subsequent faster times would hardly have been noticeable to the user. Those same users would have noticed changes to downloading photos and PC files after 2001, but then these times all remain very quick, and so at this point the difference is hardly noticeable.

Year	Upload data rate kbits	Office files (PowerPoint)		Photos		Music song	
1997	34	0.4 MB	98 sec	0.8 MB	195 sec	1.0 MB	244 sec
2001	64	0.5 MB	64 sec	0.8 MB	102 sec	2.0 MB	256 sec
2005	100	0.8 MB	66 sec	1.5 MB	123 sec	3.0 MB	246 sec
2009	500	1.6 MB	26 sec	3.6 MB	59 sec	7.0 MB	115 sec

Table 4: Historical data showing file size and response times for uploads.

Table 4, dealing with *uploading* speeds, indicates that the change of upload transmission speeds was less than for downloading, but the overall picture remains the same – file size may have increased but upload times have decreased.

Table 5 shows the typical *streaming speeds* for different applications. Streaming data from an online source to one's computer, as in watching a live video or listening to a live internet radio programme, is only possible if the network supports the required streaming speed. One may adapt the speed requirements by selecting lower quality sound or videos, e.g. lower image resolution. However, if the available transmission data rate is not sufficient, the programme is not just delayed – it cannot be viewed at all. In that case the programme might be downloaded and viewed later. Hence the user has to switch to batch processing mode.

We can take the case of YouTube to show changes over time. Table 5 shows that the videos that one could watch streamed from YouTube were always, in a sense, 'viewable' at the broadband connection speeds available at different time points. But the quality did vary. In fact, YouTube now offers high definition to take into account the fact that some viewers now have very high connection speeds, but users with lower speeds select videos at a lower quality.

Date introduced	Type of streaming media	Typical (average) data rate (kbit/s)
	Audio	
1998	Music MP3	48
2002	Internet radio	64
2005	Internet telephony (Skype)	128
2005	Audio podcasting	64

		Video	
2003		Video podcasting	495
2005		YouTube videos, standard quality, 320-240	300
2008, March		YouTube videos, medium quality, 480-360	700
2008, Nov		YouTube videos, HQ (720p)	3,500
2009, Nov		YouTube videos, HD (1080p), 1920-1180	8,000

Table 5: Streaming speeds for different streaming applications.

The experience of broadband speeds

As noted earlier there is a difference between the advertised connection speeds claimed by ISPs (also known as the 'headline speed' or the 'target speed') and the real speeds that people get in practice (also known as 'effective speed').

Over the years ISPs have regularly increased the connection speed, as shown in figure 3 for Switzerland where internet access is mainly provided by ISPs working with Digital Subscriber Line (DSL) and Cablenet Technologies. Figure 3 illustrates the historical evolution of advertised internet access data rates for the most widely used broadband product in Switzerland. We will call this product, which costs around 35 Euros a month, the 'standard product'. In Switzerland, over the last 8 years the access speed provided by the standard product roughly doubled every two years, and this was at constant price. Standard product users were upgraded automatically when the higher connection speed became available. The ISPs upgraded these access speeds in order to increase or keep their customer base – i.e. for product management and marketing reasons. This is illustrated by the fact that the two large internet providers in Switzerland typically followed each other's speed increases within months. It is worth noting that, in addition, there were new products that regularly became available which had significantly faster speeds but at higher cost. However, in practice even when the cost of these products was only 10% higher cost than the standard one, only a minority of users upgraded to the significantly higher speeds.

Figure 4 shows the average speed experienced by users in Switzerland measuring their connection in comparison to the advertised speed. The historical evolution of these speeds is shown for the most widely used (standard) product and for the fastest available product at a given time. As advertised speeds continuously get faster (see Figure 3) it becomes more difficult to achieve those advertised speeds.

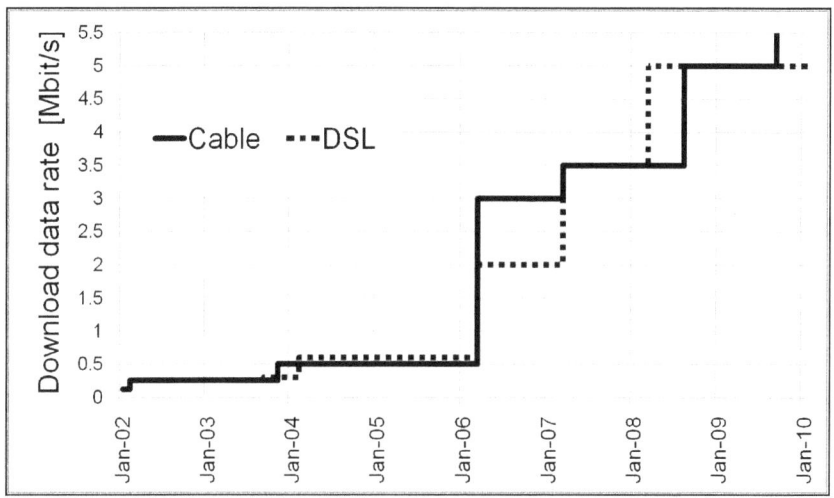

Figure 3: Evolution of advertised transmission data rate [Mbit/s] for the most widely used internet connection products.

Figure 4 illustrates that the effective speed of the standard product was about 95% and 72% of the advertised speed in January 2005 for DSL and Cablenet, respectively. At that time the standard product was at 1,200 kbit/s download data rate. However, in July 2009, when the standard product was at 5 Mbit/s the effective speed was only at 82% and 71% of the advertised speed for DSL and Cablenet, respectively.

If we start to look at speed in more detail, the experience of speeds becomes even more nuanced.

First of all one has to define the endpoints of the transmission i.e. the network connections. It makes a difference whether we look at speed for a download from the local server of one's own ISP compared to a download from a remote server somewhere else in the world. For accessing the ISP's local server we just need the so called 'access network' i.e. the network between the subscribers' homes and the internet service provider's backbone network. In access networks with Digital Subscriber Line (DSL) technology, the length of the telephone lines might be the limiting factor. In cable networks we might have local overload situations because many users share the same cable. For accessing a server somewhere in the world – perhaps on another continent – the traffic has to cross international lines going from the provider's backbone network to the final destination somewhere in the world. This traffic is typically routed over different network paths for different providers, implying different speed limitation for different providers. Finally the speed limiting factor might not be the transmission but the capacity of the server providing the information or the client software consuming the information.

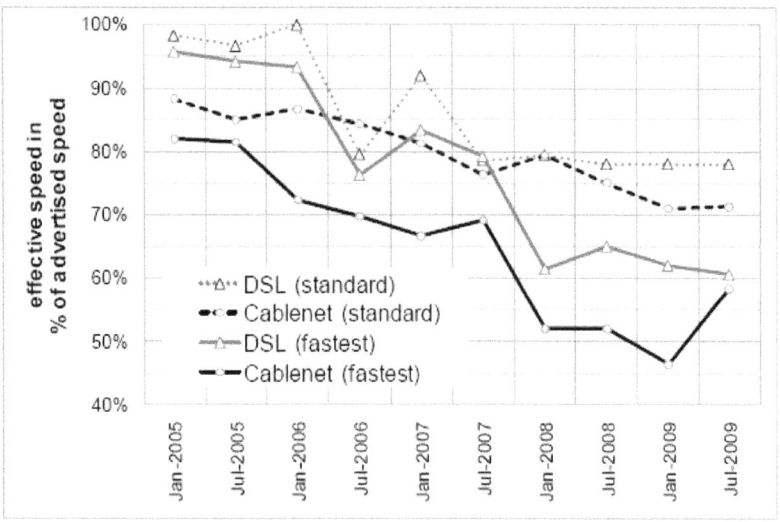

Figure 4: Evolution of measured (i.e. effective) connection speed in comparison to the advertised speed for fastest and standard (most widely used) products in Switzerland.

One important factor for the copper wire connection of DSL (as opposed to cable connections), is the length of the line from the exchange. Figure 5 shows how the maximum achievable DSL data rate decreases with increasing distance from the exchange. ADSL1 is the older, slower scheme providing speed of up to 8Mbit/s download, and ADSL2+ is the newer, faster scheme with speed up to 24Mbit/s for very short distances. VDSL is the newest scheme with speeds of up to 50Mbit/s for very short distances. The speed always declines for homes located further from the exchange (more so for ADSL2+ and VDSL).

The British regulator Ofcom collected data from 1,621 panelists who had a broadband monitoring unit connected to their router in the 30 days from October 23rd to November 22nd, 2008. This analysis shows that consumers with the most popular broadband headline speed package ('up to' 8Mbit/s) only received an average actual download speed of 3.6Mbit/s (45% of headline speed).

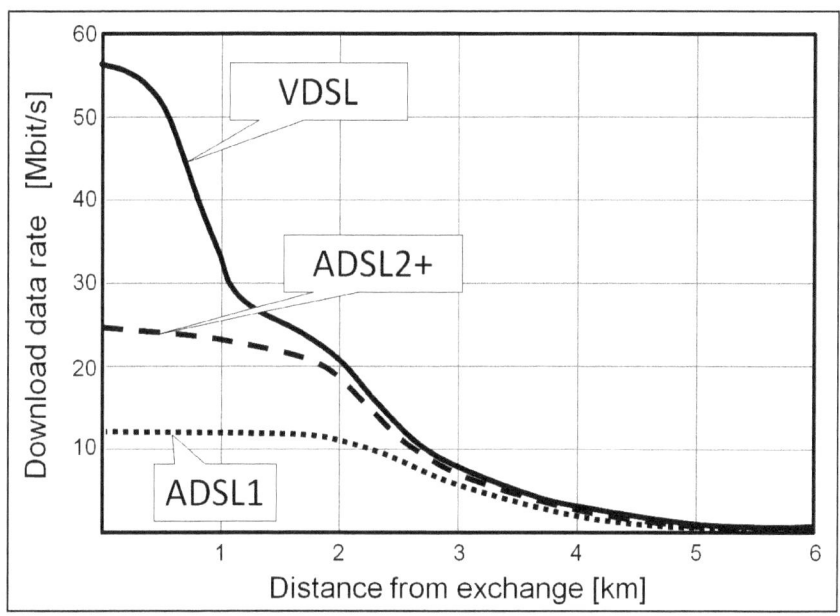

Figure 5: Theoretical maximum DSL speeds by length of line from exchange to premises Sources: Ofcom 2009, p.11 and Wikipedia).

In Switzerland similar data is collected for the most popular Swiss broadband headline speed package ('up to' 5Mbit/s) by letting users run tests in their internet browser

Figure 6 illustrates the download data rate as measured by 15,408 different consumers between April 1st and June 30, 2009. Again due to the line length, many customers only get the so called 'fall back speeds', illustrated by the numbers in the chart. Looking at these measurements in more detail, it transpires that 87% of the customers get at least 2.25Mbit/s (45% of headline speed) and 61% of the customers get at least 4,000kbit/s (80% of the headline speed). This looks better than in the UK study, but still only one third of customers really do get the advertised download data rate of 5,000kbit/s. Ofcom found speeds were 15% higher in urban areas because of typically shorter line lengths than in rural areas (Ofcom, 2009, p.33). Not surprisingly, that same study showed that rural users were more dissatisfied with their speed. That means, to take the early example, if it takes 32 seconds to download a 2 minutes video clip in urban areas where 5Mbit/s are available it would take 4 minutes in rural ones where only 600 kbit/s download speeds are available. This is a very noticeable difference, and might well make the rural user think twice about whether it was worthwhile to download.

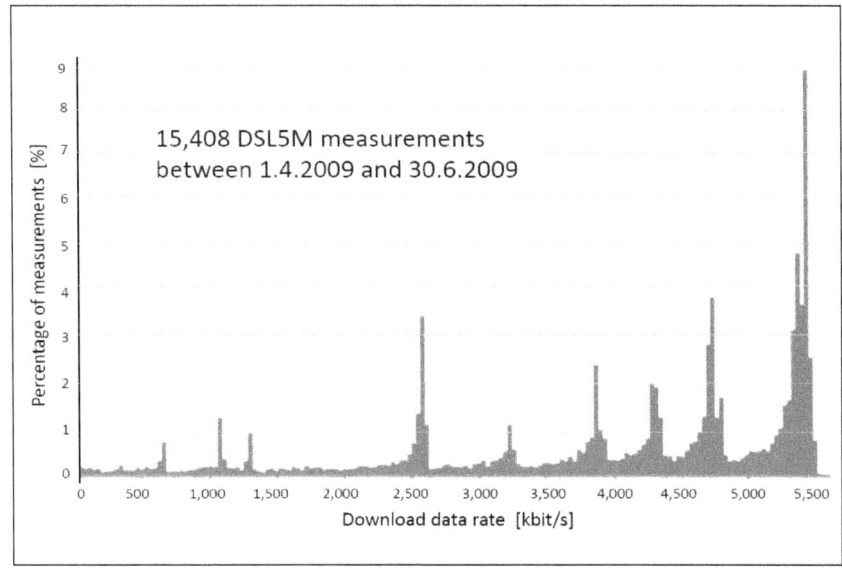

Figure 6: DSL: Experienced download data rate (in Switzerland).

Line length is not the only important factor affecting speeds. The most important factor for the Cablenet internet connections is the number and activity of subscribers who are sharing the same Cablenet infrastructure. Cablenet operators try to keep the number of concurrent users in a so called 'cell' sufficiently low. But this is always not possible in situations where there is a fast growth of subscribers or where there is a rapid growth in network use.

Previous studies of both telephone traffic and early internet use underlined the fact that people communicate or go online at certain times rather than others, and this is mainly for social reasons (De Gournay and Smoreda, 2001; Lelong and Beaudouin, 2001 – both discussed in Haddon, 2004). For example, both types of traffic can be influenced by the timing of work and of school. It can also influenced by social commitments (where there are norms about having dinner together as a family) or rules within the household (e.g. concerning when and for what purpose lone children at home can go online, for example). The point is that one of the factors affecting internet speeds is the number of users online at any one time, and so if social constraints and commitments mean that more of us go online at certain times rather than others. This creates overload situations (known as 'congestion' or 'contention') at those peak times that may make especially cablenet internet connections slower.

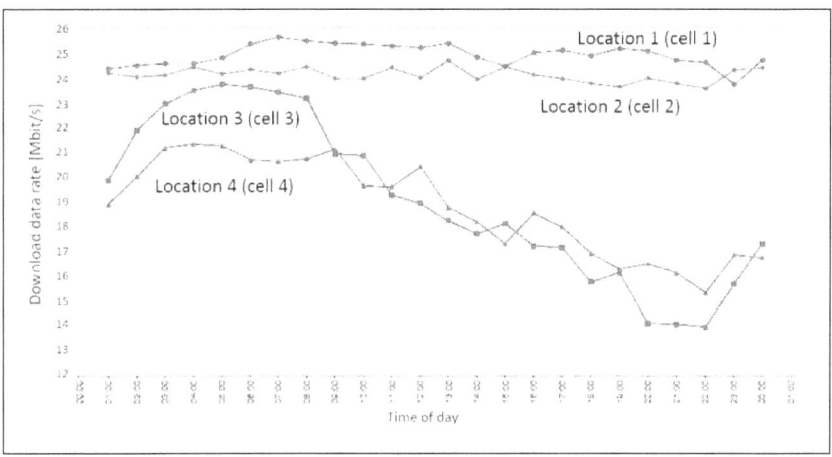

Figure 7: Cablenet: Experienced download speed by time of day (in Switzerland).

This is exactly what is shown in Figure 7 where 25Mbit/s cablenet speeds start to decline after working and school hours at certain locations – i.e. cells – while speed is fairly stable in other locations. This data from the Swiss speed tests illustrates how the effective end user data rate is influenced by concurrent users: Speed declines considerably into the evening, only starting to rise again after 10pm[53].

How much difference does this all make to user's experiences? It depends. In certain situations (i.e. cells) there is no decrease at all because there is enough network capacity available to serve all concurrent users. In other situations speed decrease of up to 50% may be noticed.

At the very fast speed of the example in Figure 7 there is only a small response time difference when downloading a 2 minutes video clip. If somebody downloads at peak load time it takes 11 seconds i.e. only 5 seconds more than it would take during the low load period.

Overload may also occur in the backbone or at the server side. In the Ofcom example from figure 8 it is shown, that there may also be a time of day dependency in the case of DSL access. In Figure 8 we notice a dip in speeds after 5pm and an increase after 11pm, where the peak hour of 5-6pm on Sunday can be

---

53 The pattern is different for the weekends (or vacation periods, such as school holidays) but the graph here is clearly strongly influenced by the figure for patterns for weekdays outside of vacations.

over 30% slower than the average speeds during the off-peak hours between 4-7am. This dip is likely to be the result of contention within ISP networks and the broader internet or at the server side, meaning that speeds are also degraded as multiple users share the same bandwidth and server resources (Ofcom, 2009, p. 28).

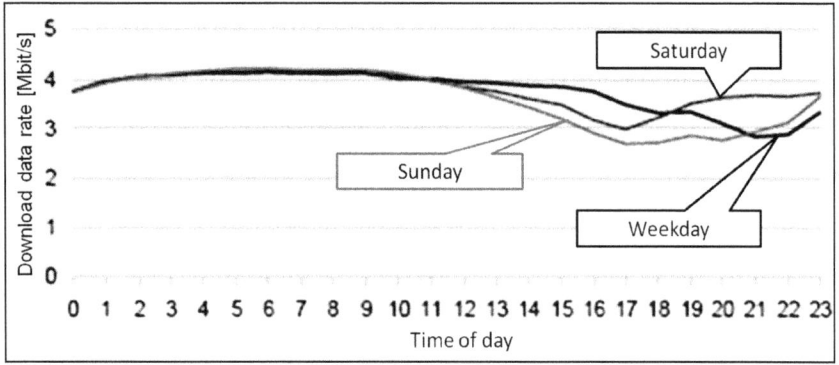

Figure 8: Average download speeds for panellists by hour of day and day of week (UK).

What does this all mean to end users? In the Ofcom example from figure 8, if they tried to download a 2 minutes video clip it would take 60 seconds at peak time instead of 38 seconds during low load periods.

Measurements in Switzerland outline speed differences that occur when connecting from an ISP backbone location to international web servers. The average download data rate from local servers (i.e. servers in Switzerland) is 82 Mbit/s, which is close to the maximum possible connection speed of the test system. In contrast, the download data rate from servers in Asia is a mere 2 Mbit/s, i.e. more than 40 times slower than the one to local servers. Looking at other continents we get 36Mbit from servers in Europe, 16 Mbit/s from servers in the USA and 4 Mbit/s from servers tested in Africa. Hence, it not only means slower speeds from international servers, but considerable variation depending upon which part of the world one is dealing with. These differences are mainly due to differences in the response time when connecting to these servers, which leads to TCP/IP protocol speed limitations. The TCP/IP protocol uses a so called receive window size parameter to control the message flow. In widely used (e.g. Windows XP) operating system's default configurations this parameter is fixed and too small for large response time and high data rate values. With novel operating systems (e.g. new Linux distributions and Windows Vista / Windows 7) or with specific TCP/IP parameter tuning, the client and the server do adapt their receive window size and therefore can achieve higher speeds at high response time.

The cross-national dimension

Figure 9 (p. 115) shows how the average download data rates evolved since 2002 in Austria, Germany, Switzerland, France and Italy based on end user measurements.

These speed differences are due to a combination of speed limits on some paths and on round trip times which lead to speed limitations imposed by the transmission protocol i.e. by TCP/IP. The speed furthermore differs by ISP due to the different paths used to access these reference servers. Looking at these historical data in Figure 9, some of the gaps in speed (here download speed) have grown wider. The Swiss and Germans, recently joined very dramatically by the French, have increasingly higher actual speeds than the Austrians and Italians – who have to wait longer when downloading standard international files, like the Skype one cited earlier. Returning to our example of downloading the 143 minutes DivX Quality Movie (i.e. 940 MBytes), if the average Swiss, German or French user enjoys a 7Mbit/s download data rate in January 2009 whilst the average Italian or Austrian user gets only 3.5Mbit/s, it will take the latter over twice as long to download the same file, 36 min instead of 16 min.

Figure 10 uses the COST298 data to compare the distribution of speed in the five countries Austria, Germany, Switzerland, France and Italy in 2009. At the point when the measurements were taken, the highest headline speed offered (called the nominal data rate on the graph) was 20,000 kbits/s (or 20 Mbit/s). The 'data points' at the bottom refer to the numbers in the various national samples, i.e. the number of times people checked their speeds with the speed test applications. The number of measurements carried out is higher since in practice most check their connection speed three to five times. Hence, 10% of this, (the 10 on the horizontal axis), was 2 Mbits/s, and 65% of the Austrians, French and Italians had at least this speed, 75% of Germans had a least this speed and 85% of the Swiss had a least this speed. Country differences exist at this end of the scale, but as we look at even slower speeds, going to the left in the graph, country differences diminish since most people in more of the countries can have at least these speeds. If we now go to the right in the graph and look at 50% of the nominal data rate, 10 Mbit/s, only 5% of Austrians and Italians have this speed or greater than it, whilst the figure for France is just over 15%, and for Germany and Switzerland 30%. This conveys some idea of the level of country variation and reminds us that before we consider averages the reality is that we are always talking about proportions of the internet users with different speeds.

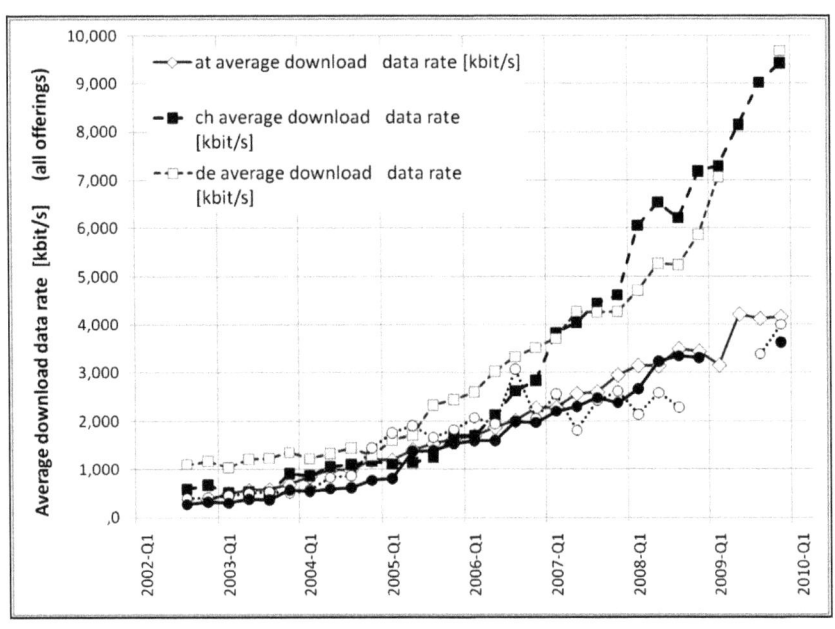

Figure 9: Evolution of experienced download rates by country.

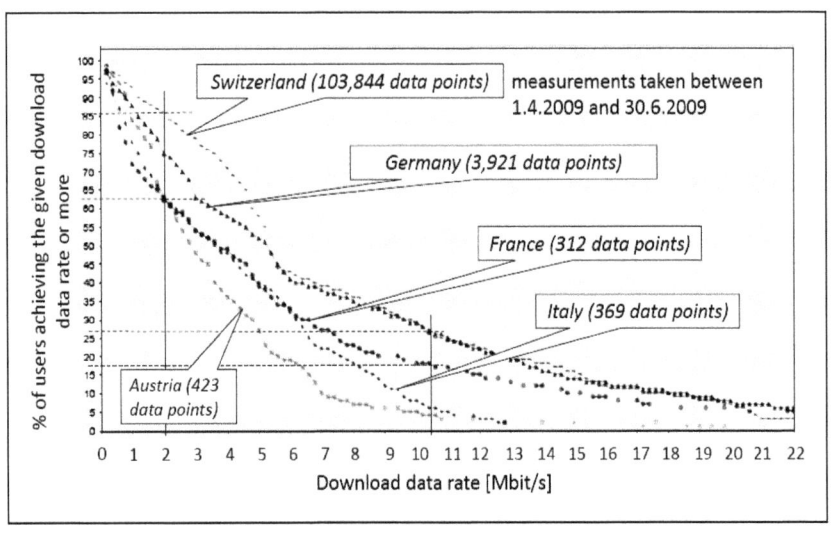

Figure 10: Distribution of experienced download speed in different countries.

Conclusions

In general, this chapter has been an interdisciplinary exercise, unpacking the technical considerations to be considered and translating this into terms that connect with some of the more social science based approaches and questions about the nature of internet use. Moreover, it has attempted to achieve this through comparative analyses both within and between countries, an approach that could be used in subsequent research.

The first substantive point to observe is that the historical data showing the growth in broadband speed and file sizes illustrate the interdependence of the content offered and the available download speed. As soon as higher speeds are broadly available the application designers begin to offer better quality, e.g. resolutions in the packages involving image, and video files of longer duration. On the other hand, availability of high volume content then in turn drives the evolution of access speed. The implication is that this is an on-going process. In other words, speed was not only an issue in the early years of the internet but it will always be an issue because there will always be some new applications 'eating up' all bandwidth.

Certainly the online distribution of software packages, updates and service packs would not be possible at the speeds we had ten years ago. Some things have been made possible, or made more practical, by the faster internet. But while on an everyday basis many of the tasks people conduct can be achieved in reasonable time, history suggests that applications will be developed that in the future require more of the public to upgrade to yet faster connections. The currently much discussed application service-providing solutions (aka "cloud computing") require speeds of at least 15Mbit/s to allow for sub-second response time when working with current PowerPoint files of about 1.6MB file size.

These histories also provide us with a good idea of what will happen in the mobile world in the next few years to come. The speeds that we had on the fixed internet connections ten years ago are now common on mobile phone connections. It is expected that the evolution of mobile phone data rates will be even faster than it was on the fixed internet. But the interdependence of content offered and available download speeds will be very similar.

The next question addressed in this chapter is that of what these changes in speed and file size mean to the user, that is, how are they experienced? This required some technical clarification, since the average size (at any one time) of different types of file, types of video, and types of photo will vary. Nevertheless, the tables provide some sense of the history of when (and at what connection speeds) an increase in speed was noticeable to a user and for what types of application. This becomes important because while social science research regularly asks people (such as the Italian youth cited earlier) what they do online, the technical analysis here conveys a sense of how much time it requires to achieve different goals (including the quality of what can be viewed) – for example, the time taken for peers to acquire, exchange or post online audio-visual material.

The comparative analysis within countries takes this a stage further. When we hear that a certain speed is available for companies or people in certain areas, not all those people have it. The analysis then goes on to show how much where you live can make a difference to the speed experienced, particularly relevant for rural vs. urban differences. Moreover, the time of day when you go online also makes a difference, and in some cases their other commitments means that certain people may only go online at more congested times. All these factors highlight the ways in which the experience of using the internet can vary, certainly in terms of the time taken and hence attractiveness of particular actions.

Finally, the comparative analysis conducted cross-nationally illustrates not just the gaps between countries (in terms of the average experience of speeds) but also how these gaps change over time, with certain gaps expanding of narrowing, sometimes slowly, sometimes more dramatically. As before, translating these into the time that users in different countries might take to achieve certain goals castes some light on why practices are more attractive in some countries more than in others. While other social factors influence patterns of use, the technical constraints noted in this chapter can also play a part.

Bibliography

Agrawal, N. Bolosky, W.J., Douceur J.R., & Lorch, J.R., 2007. A five-year study of file-system metadata. *ACM Transactions on Storage*, 3 (3), Article 9.

Bakardjieva, M., 2001. Becoming a domestic internet user. *e-Usages*. Paris, France 12-14 June 2001. ADERA: Bordeaux.

De Gournay & Smoreda, Z., 2001. La sociabilité téléphonique et son ancrage spatio-temporel. *e-Usages*. Paris 12-14 June 2001. ADERA: Bordeaux.

Growth of the Average Top 1000 Home Page, 2008.

Haddon, L., 2004. *Information and communication technologies in everyday life: a concise introduction and research guide*. Oxford: Berg.

Haddon, L. & Heinzmann, P., 2009. *The Experience of Broadband Speeds*. A report for COST298. [Online] Available at: http://www.cost298.org/uploadi/editor/1264711174Changing%20broadband%20speeds%20report.doc [Accessed 1 March 2010].

Haddon, L. & Kim, S-D., 2007. Mobile phones and web-based social networking - emerging practices in Korea with Cyworld. *The Journal of the Communications Network*, 6 (1), January-March.

Lelong, B & Beaudouin, V., 2001. 'Usages d'internet, nouveaux terminaux et hauts debits: premier bilan après quatre années d'expérimentations. *e-Usages*. Paris 12-14 June 2001. ADERA: Bordeaux.

Ofcom 2008. UK broadband speeds 2008. Consumer experience of broadband performance: initial findings. [Online] Available at: http://www.ofcom.org.uk/research/telecoms/reports/bbspeed_jan09/ [Accessed 19 July 2009].

Ofcom 2009. Accessing the internet at home: a quantitative and qualitative study among people without the internet at home. [Online] Available at: http://www.ofcom.org.uk/research/telecoms/reports/bbresearch/bbathome.pdf [Accessed 19 July 2009].

Vesna Dolničar, Charalambos Christou, Rosemarie Gannon, Leslie Haddon, Pedro Puga and Jorge Vieira

Chapter six. Cross-national broadband digital divides

Introduction

The digital divide debates, originally focused on the era when the internet was narrowband, provided a comparative analysis in both senses considered in this book: within societies and cross-nationally. The aim of this chapter is to understand emerging digital divides in the era of broadband and more contemporary services and practices online.

One theme from the digital divide literature was the argument that it was based on existing, offline inequalities and was often replicating those divisions online, entailing relative advantage for some and disadvantage for others. While many of the commentaries about the contemporary internet, especially the web 2.0 services discussed in earlier chapters, are very positive it is nevertheless important to ask if new divisions are emerging within internet users: between narrowband and broadband. This becomes particularly salient in the light of claims that broadband enables richer experience, activities, and practices. Hence one goal of this chapter is to explore the extent of any such gaps (and indeed whether broadband experiences appear to be richer).

The second goal is to introduce the cross-national comparative perspective. Over the years quite complex society measures of broadband, indeed indices of the experience of broadband, have been developed enabling us to make more nuanced distinctions as regards any digital divide between countries. But this comparative perspective may also provide further clues about the processes leading to some of those narrowband and broadband differences.

The rest of the chapter proceeds as follows. It introduces some of the original digital divide debates. It then reflects on some of the claims made about the experience of broadband, especially web 2.0 services. It outlines some of the methodological considerations in the analysis that follows. And it provides a secondary analysis of Eurostat data of selected services in five European countries: Cyprus, Ireland, Portugal, Slovenia and the UK.

Digital divides

On the one hand the promise of the internet – web use, for example – means tapping new 'digital opportunities' to support the inclusion of socially disadvantaged people. As Hargittai (2008) points out, from the perspective of social mobility, digital media could offer people, organisations and societies the opportunity to improve their positions regardless of existing constraints. Yet the potential of ICT to overcome social exclusion, enabling the growing social inclusion of individuals, should not be exaggerated. From the early days of the digital divide debates, Menou (2001), van Dijk (1999) and Norris (2001) maintained that – from the point of view of social reproduction – the explosive growth of the internet actually exacerbated existing inequalities between the information-rich and information-poor, both within certain countries and between them. In contrast to the more optimistic views noted earlier, various uses of the internet thus have the potential to increase the inequalities that result from the accumulation of advantages provided by wealth:

> "benefiting those who are already in the advantageous positions and denying access to better resources to the unprivileged." (Hargittai, 2008, p. 943).

Merton's (1973) identification of the Matthew effect – 'Unto every one who hath shall be given, and he shall have abundance' – applies to this position.

Irrespective of these contradictory understandings of the phenomenon in academia, a consensus nevertheless exists that the concept of the digital divide is somehow tightly linked to the concept of social inequality (e.g. Attewell, 2001; Bonfadelli, 2002; DiMaggio, et al., 2004; Mason & Hacker, 2003; Menou, 2001; van Dijk, 2005; Warschauer, 2003). Parayil's (2005) reasoning that the digital divide presents both processes, being a symptom and a cause of broader social and economic inequality, seems most reasonable. Hence, for the purposes of this chapter, the digital divide is both seen as a result of social exclusion (those who suffer from a lack of financial resources, skills or capabilities will have trouble accessing ICTs and handling information that is accessible through ICTs) and as a factor that will aggravate the other dimensions of social exclusion.

Broadband and new practices

There is a claim that the diffusion of broadband has stimulated more interactive and participative uses of the web, encouraging users to become more creative content producers (e.g. Ewing & Thomas, 2008; Tolbert & Mossberger, 2006).

In support of this, various empirical studies have suggested that the affordability of broadband is enabling changes in user behaviour. One OECD study showed that:

> "New content-rich broadband applications and new forms of usage have become a key driver of broadband demand and uptake. The availability of broadband has reinforced existing activities (e.g. e-mail, news and information, shopping online), but this has also brought about new forms of usage and innovation (e.g. video streaming, podcasts, high-definition television over broadband)." (OECD, 2008, p. 89).

A related study showed that broadband users tended to contribute more content to websites, keep online diaries and blogs, and share photos, videos and artwork (OECD, 2007). Similarly, an Australian World Internet Project study (Ewing & Thomas, 2008) revealed a close positive relationship between the diffusion of broadband services and creative uses of the technology, where users of broadband were more likely to post videos and photos, to download music and to listen to podcasts and radio online, for example.

But the secondary data analysis conducted for this chapter also suggests that different components of broadband may be more important for different activities; while in some activities the 'always-on' feature is more relevant, for others speed comes more into play.

Measuring the broadband experience over time and cross-nationally

As examined in some depth in the chapter five, the fast changing ICT environment and constant new technological achievements and developments demand the continuous (re)changing of our understanding of broadband. The term broadband is today used to describe almost any always on, high speed connection to the internet. Broadband is often associated with a particular speed or set of services, but in reality the term 'broadband' is like a moving target, with internet access speeds increasing all the time.

What is it about current broadband that makes a difference to the broadband experience? The recent study of the quality of broadband connections experienced by individuals in 42 countries (in Europe, North America, OECD and BRICs – Brazil, Russia, India and China)[54] included several performance parameters, which were grouped into three major categories: download and up-load throughput and latency. These factors appeared to change substantially the way

---

54 The press release of the study is available online at: http://www.saeurope.com/newscenter/downloads/CiscoBroadbandQualityrelease.pdf [Accessed 26 February 2009]

in which the internet is used in domestic environments, opening up new possibilities for what may be done online. However, one could expect that not all features of broadband are (equally) important for specific online activities. While in the case of some activities high bandwidth may be very influential, in case of others the speed of the narrowband might be sufficient, but the impact of the always on aspect of broadband makes the involvement in specific web uses more appealing.

Turning now to the issues involved in comparing broadband cross-nationally, the broadband gap between countries should not be based on mere access measured by penetration rates, but should include quality and capacity divides, as well as differences on the range of services people can access and use. In the above mentioned study of the quality of broadband connections, for example, the Broadband Quality Score (BQS) for each country was determined using a formula that weighted each category according to the quality requirements of a set of popular applications available now and in the future. Typical applications for today include web browsing, social networking, music downloads, basic video streaming and video chatting, standard definition IPTV (the internet on the TV), and enterprise-class home offices. But we must remember that, once again, the very components of these indices may evolve. Future applications that may feed into countries' scores for broadband quality may include consumer telepresence for communications, healthcare and education, high-quality video file sharing and streaming, high-definition IPTV, cinema-quality live event broadcasts and advanced home automation.

Finally, there are the parameters of the particular secondary analysis conducted for the chapter. We are interested in the potentially richer online experiences brought about by the content-rich broadband applications, but at this point we are limited by the secondary data that are available. So we can only present the data regarding the number of different activities that individuals participate in online (comparing narrowband and broadband). Since we are confined to the data available in the Eurostat database it is also not possible to examine the relationship between, for example, e-skills and broadband availability. Our empirical results thus deal exclusively with the indicators that measure online activities related to communication, creation of content, peer-to-peer file sharing practices and software downloading. As evident from the results that follow, it is valuable to explore all the activities mentioned, because each of these practices can be contextualised nationally.

Regarding the time range of the data, the change examines only the latest available data, i.e. 2007. Even though the earlier time series data are important, there is a significant amount of potential data and this could bring about information overload. In addition to this, our interest lies more in revealing cross-country and within-country asymmetries than in the way that they evolve over time.

Cross-national broadband divides

Let us first introduce a brief overview of the infrastructure conditions and level of access by households in each of the countries (Cyprus, Ireland, Portugal, Slovenia and the UK), dichotomised between narrowband and broadband in Figure 1.

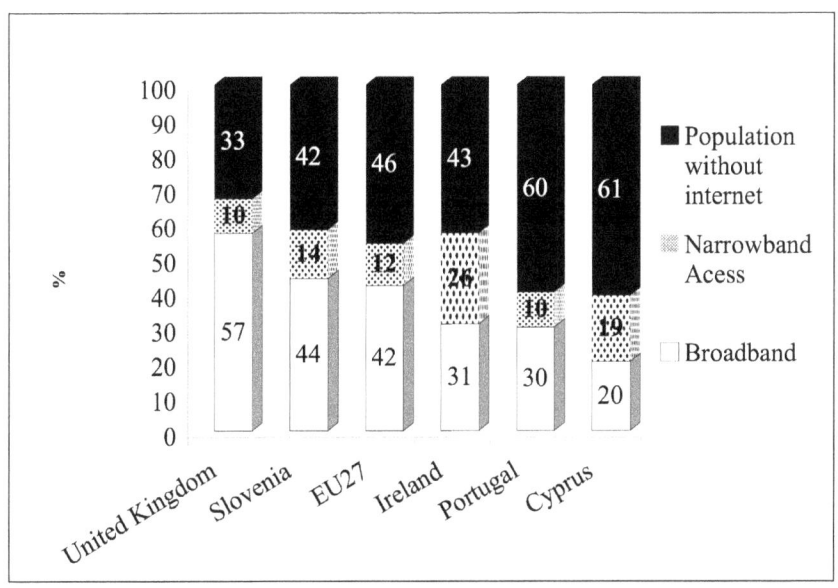

Figure 1: Internet access by households, 2007 (% based on total national population).
Source: Eurostat Base: 16-74 year olds

Of all the countries studied, the United Kingdom shows the highest level of overall internet access and broadband penetration rate. Slovenia comes in second place and Ireland in third place. However, a better ratio of broadband versus narrowband access exists in Slovenia. Portugal occupies the fourth position, closely followed by Cyprus; although, Portugal shows a superior share of broadband households.

The issue of cross-national differences can be explored in more depth using the Broadband Performance Index (BPI) (Commission staff working document, 2008). The BPI seeks to compare the performance of broadband in member states according to a range of factors: broadband coverage (reflecting developments in rural areas), competition by coverage (reflecting countries' innovative capacity, propensity to invest and consumer choice); speeds (reflecting quality developments), prices (reflecting affordability), use of advanced services (reflecting the propensity of individuals and businesses to take up innovative services and the perception of trust), and socio-economic context (reflecting preferences, skills and capital equipment that influence the propensity to use advanced communication technologies and services).

The above work concludes that as regards the countries under study the UK performs well according to most indicators but lags behind in terms of speed and take-up of advanced services, in particular by businesses (although trust in the online environment is generally positive). Portugal and Slovenia have a weaker 'socio-economic context' (see above), in particular in terms of ICT expenditure and skills, with limited use of advanced services as well as relatively high prices. Cyprus' and Ireland's performances are limited along most dimensions by the socio-economic context. Cyprus is also limited by high prices but in contrast has good broadband coverage in rural areas.

So this more nuanced approach to cross-national comparisons shows us that it is important to take into account not only the basic penetration rates (reflected merely by the percentage of households that have broadband access), but also other factors, that indicate the qualitative aspects of the internet experience. By considering these factors one can avoid drawing one-sided and partial conclusions about the role of the internet in specific contextual environments. Let us take the Slovenian example to illustrate this: while Slovenia shows high internet coverage (as depicted in figure 1), the abovementioned Broadband Performance Index reveals that Slovenia performs worse in most of the broadband attributes, particularly those related to more skilled and advanced use of broadband services and applications (which is due in particular to trust-related indicators).

These (mostly structural-level) factors that have an impact on the take-up of broadband services in general are also very useful for our further analysis of people's involvement in specific online activities within different countries.

Broadband versus narrowband online practices

At the most general level, our data analysis points to a positive correlation between the availability of broadband at home and people's engagement in a variety of online activities. Within the observed countries, on average approximately 80% of individuals with a broadband connection use the internet regularly, while only around 60% with a dial-up connection do so (Commission staff working document, 2007). Similarly, citizens in countries with the highest broadband penetration rates make more intensive use of high bandwidth demanding services.

When we turn to specific services it is useful to establish some of the principles of analysis with older applications before looking at the newer ones. Therefore, the first specific application we review is one of the most basic and long established uses of the internet: sending an email with attached files. Since the majority of these short messages usually only require limited bandwidth, it is not surprising that we find only minor differences between narrowband and broadband users (Figure 2[55]).

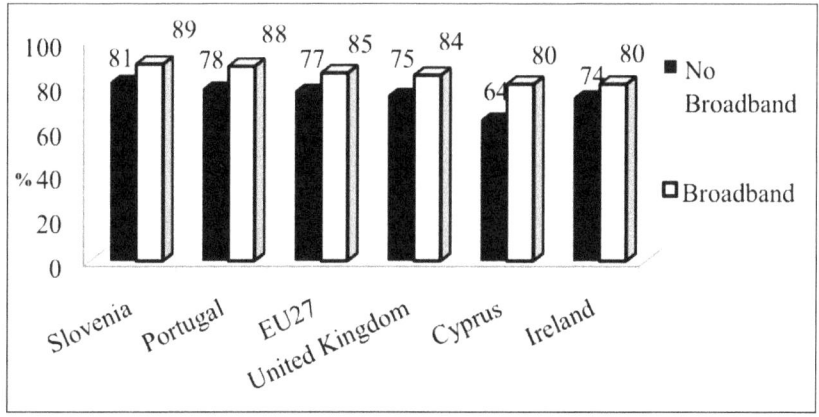

Figure 2: Percentage of internet users who have sent an email with attached files, 2007. Source: Eurostat Base: 16-74 year olds

As will become clear later, there is always some variation between the narrow versus broadband gaps in different countries, but here the range is relatively small, 6-14% (the latter in Cyprus). The small gap between broadband users and narrowband users might be explained by some technical differences i.e. the fact that sending emails with (larger) attachments using a broadband connection might be (slightly) faster. But it may be due to the fact that broadband users have a different profile from narrowband ones, and simply send slightly more emails because of that. Lastly, the overall differences (across the countries considered) between the percentages of broadband users that send an email with attached files are generally small. The relevant percentages range between 80%

---

55 The data illustrated in Figures 2-7 relate to the percentage of individuals who have ever used the internet. The category 'Broadband' indicates individuals living in a household with broadband access and category. 'No broadband' indicates individuals living in a household with internet access but with no broadband access. The population considered is aged 16 to 74 years.

and 89%, as observed in Figure 2. The small differences can be attributed to the fact that sending emails with attachments is a widespread practice among internet users.

The next step is to look at the less widespread but still established textual communicative practices – posting messages to chat rooms, newsgroups and online discussion forums. This is generally more relevant to discussions of whether the internet leads to more forms of social participation since these all count as various forms of online participation.

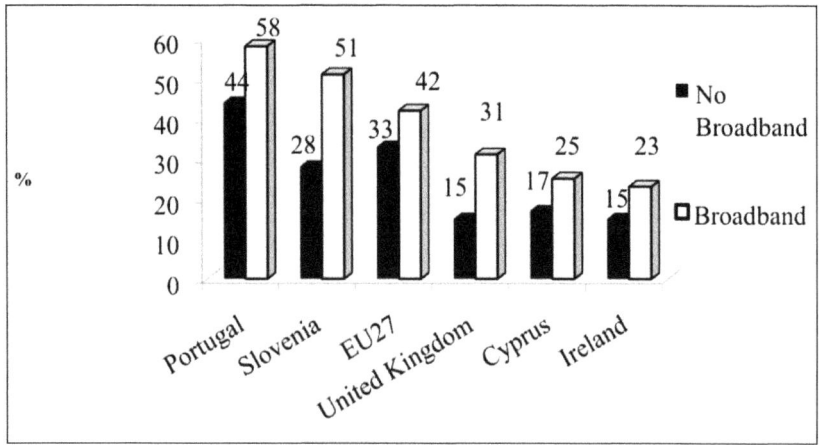

Figure 3: Percentage of internet users who have posted messages to chat rooms, newsgroups or an online discussion forum, 2007.
Source: Eurostat Base: 16-74 year olds

In Figure 3 we can see both a gap between narrow versus broadband users and also differences between countries – and both are greater than in the case of email. This is illustrated in the case of the cross-country divide by the fact that half of broadband internet users in Portugal have posted messages as opposed to less than a quarter in Ireland. Moreover, there are national differences whether one compares broadband or narrowband, so there must be some social reasons for national variation in these practices.

As regards the technological differences in terms of posting messages to chat rooms, newsgroups and online discussion forums, in all the countries, the differences between narrowband and broadband are much more dramatic than in case of sending emails, and the range of gaps is greater than for email, being largest in Slovenia (23%) and smallest in Cyprus (8%). As in the case of email, the bandwidth should make very little difference to the speed of posting such short messages. There might still be some technical influences on this pattern, such as the always-on connectivity associated with broadband. But once again, it may be that the type of people who adopt broadband are ones more likely to participate online – in this case, the difference between the profile of narrowband and broadband users might be larger in Slovenia and smaller in Cyprus.

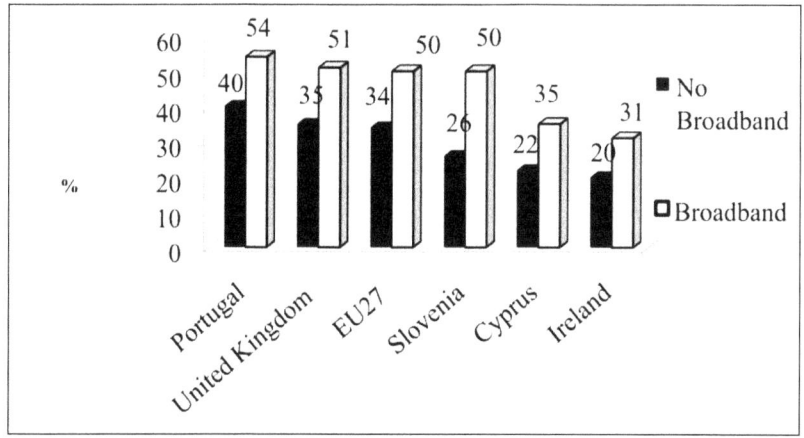

Figure 4: Percentage of internet users who have downloaded and installed software, 2007. Source: Eurostat Base: 16-74 year olds

Before considering the newer services, it is useful to think about the ways in which we know how older services have been affected by broadband. Downloading software is a long established practice, pre-dating web 2.0. But in Figure 4 we see that those living in a household with broadband access are also more likely to have found, downloaded and installed software. This is understandable in light of the growth of software file sizes, as shown in chapter five, which almost assume broadband access – i.e. it would take a long time to download them with narrowband. In other words, the act of downloading may not be new, but changes in file sizes have an impact on the nature of this process, making broadband significant. That said, some narrowband users say they download software, although this in part may reflect the fact that they may be referring to software downloading in the past when files were smaller.

As in the previous services, the gap (including broadband downloading) between countries, especially Cyprus and Ireland versus the other countries, suggest some social factors influencing this practice. These factors may be related to those affecting online participation, since once again, Portugal is at the top and Ireland at the bottom.

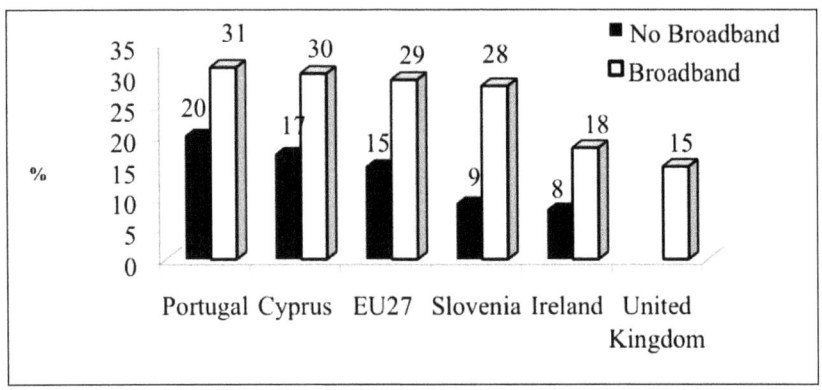

*Figure 5: Percentage of internet users who have used the internet to make phone calls, 2007.
Source: Eurostat Base: 16-74 year olds*

Having established that differences exist between countries and narrowband versus broadband users even in the case of older practices, we can now consider the newer ones. We start with verbal communications online (e.g. via Skype), where we expect broadband to make a difference because high bandwidth is of great importance for making phone calls. Figure 5 shows that in all countries, those who have broadband access are more likely to have made internet phone calls. In all countries where we have data available the gap is quite large. Looking across countries, Portugal and Cyprus have the highest rates for both narrowband and broadband access while the United Kingdom has the smallest proportion of individuals who have used the internet to make phone calls[56]. But we can introduce a new element of cross-national analysis here – the two countries otherwise lagging in terms of internet (and also specifically broadband) penetration (as illustrated in Figure 1) rank among the top countries. In contrast, the UK, the otherwise leading country in terms of internet adoption rates, is last. Moreover, in the case of services where broadband should make a difference, variation in the use of these services is also influenced by cultural, social and economic issues and not just by the broadband penetration rate in a particular country. It might be, for example, that in Portugal and Cyprus (non-internet) phone calls are more expensive (in absolute and/or relative terms).

---

56 There are no UK data for those with no broadband access.

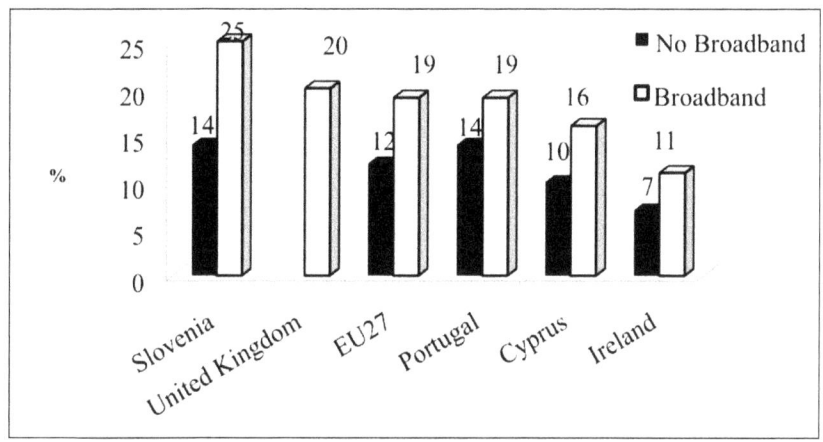

Figure 6: Percentage of internet users who have created a web page, 2007.
Source: Eurostat Base: 16-74 year olds

We can now look at some potential creative practices associated with web 2.0. As in the case of downloading software, there are in practice continuities between old and new applications in this field. It has long been possible to create web pages, even with narrowband. But the development of social networking sites (SNSs) over recent years made a difference, leading to far more web pages: for example, all the profiles on Facebook and MySpace. While SNSs do not require broadband per se, always on connectivity makes it easier to constantly check other people's online profiles and potentially contribute to these sites while high bandwidth is useful for supporting audio-visual elements (e.g. music, animations) - features admittedly used by a minority.

So does broadband make a difference? At face value, yes, since in figure 6 one can clearly distinguish a gap in all countries between those living in households with broadband access and those with no broadband access. But is this entirely caused by the nature of broadband as a form of access? To develop our arguments about the profiles between narrowband and broadband users, we know that SNSs are most popular with youth and young adults, and these are the ones most likely to have broadband – so once again, the factor causing the difference in the figures may well relate to the type of people using broadband.

Looking across countries, the largest difference within countries occurs in the one that also shows the highest percentage of web page creators, Slovenia, while Ireland has the smallest difference but also the lowest percentage of web page development. Lastly, the same principle occurs as in the case of internet phone calls – amongst the five countries, Slovenia is not foremost in broadband penetration nor Ireland last (see Figure 1), so social factors and not just technological development must be playing a role.

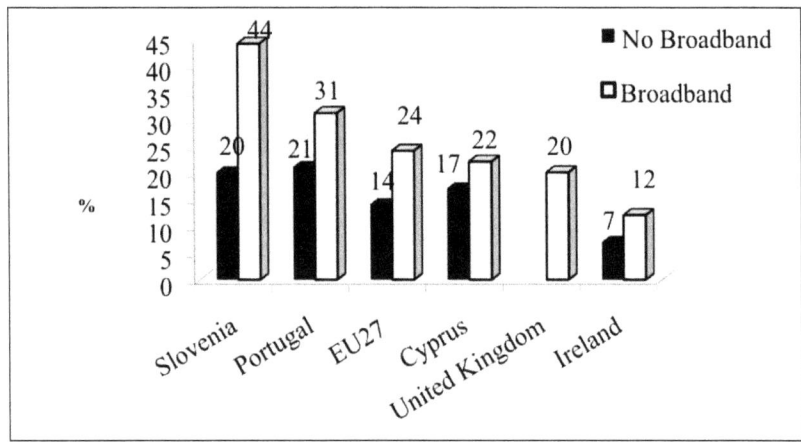

Figure 7: Percentage of internet users who have used peer-to-peer file sharing for exchanging movies, music, etc., 2007.
Source: Eurostat Base: 16-74 year olds

Finally, an important emerging internet use is online media sharing practices, enabled by peer-to-peer sharing sites (and software). As demonstrated in chapter five in this book, this benefits from high bandwidth because of the size of the files, and indeed in Figure 7 we find that more broadband users participate in this practice in all countries where data is available. But 'benefits' is clearly the appropriate word, since it can be managed with narrowband if one is willing to wait, and indeed Figure 7 also shows that there are narrowband internet users willing to do this.

Looking cross-nationally, the data shows that Slovenia has the highest rate among those with broadband access, followed by Portugal while Ireland has the lowest rate. As in the case of previous figures, this underlines the importance of social factors at work – in particular the case of Portugal with low broadband penetration, while the UK with the highest penetration rate is towards the bottom on this graph.

The above data analyses point to a relation between the availability of broadband at home and people's engagement in multiple online activities. It is clear that broadband is not only correlated with an increase in the share of the population conducting different online activities, but also relates to more frequent and longer internet use. As regards basic internet activities (like sending emails with an attachment) one can note minor differences between narrowband and broadband users; but in activities related to applications that work best with more bandwidth (e.g. peer-to-peer file sharing) the gap is wider.

Discussion

It is important to start with a qualification. At the moment, the statistics that are available show us correlations, but that cannot prove causality. To take the example of sending emails, we suggested that the correlation with broadband use might exist because the latter enables one to send larger attachments more quickly. But it may be the case that people who were previously more active on narrowband are the ones more likely to adopt broadband in the first place. Being active might have involved more communication, hence they are the type of people who send more emails. This same logic could apply to all the indicators studied. Indeed, we can imagine that there were differences in the degrees of activity between users of narrowband, and the migration of more active users to broadband has made that distinction more visible in the figures that compare narrowband and broadband.

However, from various studies, as well as our knowledge of technological affordability we also know that it is probable that some of the difference between narrow and broadband arises because the latter technology makes certain applications easy or in some cases viable (e.g. telephone conversations via the internet). Hence we suspect that while several processes may be simultaneously at work, broadband may in part have a bearing upon internet user choices and practices. In sum, while these potential processes have been emphasised, we recognise that the picture behind the statistics may be more complex and deserves further investigation than can be managed in this chapter using the data available.

Based on the data studied one could conclude that the diffusion of broadband is correlated to more interactive and creative uses of the internet, including the development and sharing of creative content. In fact, these findings are in line with other empirical research conclusions, most of which have suggested that the affordability of broadband is enabling change in user behaviour (i.e. ignoring the potential complications about causality noted above).

In terms of differences across the five countries, these are more evident in some activities more so than others. In general, Cyprus and Ireland show the lowest online participation regarding the reported activities. In Ireland many people still prefer to have face-to-face contact with others rather than meeting (new) people online. Hence, in their free time, people living in Ireland prefer to meet others face-to-face and socialise offline. This could in part explain why some people living in Ireland use the internet less for recreational purposes, i.e.

exchanging movies and music online or to post messages to chat rooms (see Figure 3), in comparison to the other countries studied. It is somewhat surprising that individuals in Slovenia and Portugal (which are not countries with the highest broadband internet penetration) are more involved in many of the activities investigated. This is particularly clear in case of two of the activities that arguably require more time and effort, i.e. peer-to-peer file sharing and sharing thoughts by posting messages within different online environments. These differences might be explained by referring to a wide variety of (and complex relationships among) social, cultural, political and economic factors.

The way in which (local) communities shape access to ICTs as they respond to changing policies, market dynamics, technological advances, and issues of social exclusion in the digital age may vary across different countries. While investigation of these determinants is not the focus of this chapter, we will nevertheless turn to some of these with the purpose of showing how these factors can help our attempts to understand at least some of the (cross-and within-country) disparities revealed by the findings of the secondary analysis presented above.

In the example of making phone calls via the internet (see Figure 5), it is evident in terms of this activity that Portugal has the highest rate and the UK the lowest rate. From one point of view, this could be seen as a surprising finding because (considering economic factors) previous UK research involving day-in-the-life-of diaries and interviews, for example, suggests that although it was by no means the only factor, the costs involved in using different channels of communication were a major consideration shaping British people's choices between channels (Haddon and Vincent, 2005). In a similar vein, Telecom Italia survey data relating to Britain (but also France, Germany, Italy and Spain) suggested there was widespread sensitivity to the price of telecoms costs in general and that this was a constraint influencing calling decisions (Haddon, 2004)[57].

Furthermore, due to the UK's diverse population (official statistics show that immigration has contributed to half of Britain's population growth in the last 10 years) it might be expected that immigrants would be attracted to free-of-charge international calls. For example, ethnographic research on West Indian immigrant communities and their use of ICTs, as well as of immigrants' communications in the Netherlands and in France (Calogirou & Andren, 1997 in Thomas, et al., 2005) studies show that ethnic background strongly influences the intensity, social composition and geographical reach of social networks and

---

57 Here it should be noted that while the cost structures provide the 'supply' part of the equation, the willingness of people to pay different costs, in the language of economics their 'marginal utilities', may nonetheless be influenced by a range of cultural factors, by the circumstances of people's daily lives and by the values related to them (Haddon, 2004).

how these networks are maintained over distance by use of the telephone, mobile phone or e-mail.

However, it seems that the low interest by the British for making (free) internet phone calls could be at least partially explained by the technological framework conditions in terms of the quality of the broadband services and infrastructure in the UK. Namely, both of the above mentioned indices – i.e. Broadband Quality Score (BQS) and Broadband Performance Index (BPI) show that the UK lags behind in terms of broadband quality. According to the BQS, for example, the UK "does not have good enough broadband to consistently deliver high quality web access, falling behind not only the usual leaders, but coming in the bottom half of ranking"[58]. Since internet phone service is a typical data intensive web application, the smaller interest in this online activity in the UK might be also a result of lower broadband quality.

Another indication of the impact of technological infrastructure (in terms of the quality of both narrowband and broadband) on different internet uses is that the greatest difference between those with narrowband and those with broadband access is seen – for almost all the cited activities and particularly for those that require high bandwidth – among Slovenian internet users. This finding could be explained by the fact that (even though broadband penetration rate is still relatively low) good latency and speeds gives Slovenia a high BQS (it is ranked as tenth among the 42 countries). Namely, solid broadband connections might encourage individuals to get more involved in (and benefit from) the content-rich internet services and applications.

Conclusions

The point is that a cross-country perspective can tell us more than the fact that there are international digital divides – which has been the focus of previous studies. Taking a cross-cultural perspective can also add to our understanding of the processes shaping digital divides within countries, showing that some processes are not universal, but strongly dependent on the different country-specific conditions. Within this context, one of the paths of our future research could follow the social construction of technology (SCOT) approach, which provides a helpful framework to explore the diffusion of high-speed, advanced communications (Dutton, et al., 2004) and is concerned with the ways in which social forces impact the design, adoption, and uses of technology (see e.g. Lievrouw, 2002).

---

58 See http://www.itpro.co.uk/606138/uk-lags-in-broadband-quality [Accessed 26 February 2009]

Possible future research could also investigate in more depth how other user characteristics, their social environments (including social networks) and online abilities influence the types of uses and also how different uses might have significant implications for (improved or impeded) life outcomes. We should note here that the phrase 'social inclusion' captures this sense of avoiding social isolation. The implication is that we need to consider not just what we possess but also what we can do, the extent to which we can fulfil various social roles – or be constrained in doing so. Approaching the role that ICTs play in relation to social inclusion in this way would enable us to explore in more detail how ongoing developments in the nature of the internet, and by extension the newer and future services and facilities online, can not only facilitate connection with the wider society but also enhance certain divisions (Haddon, 2000). Future research and policy making agendas should address this issue of new kinds of (in)equalities, arising from the various uses of ICT.

Bibliography

Attewell, P., 2001. The first and the second digital divides. *Sociology of Education*, 74, pp. 252-259.

Bonfadelli, H., 2002. The internet and knowledge gaps – a theoretical and empirical investigation. *European Journal of Communication*, 17 (1), pp. 65-84.

Commission staff working document, 2007. Accompanying document to the communication from the Commission to the European Parliament, the Council, the European Economic and Social Committee and the Committee of the Regions. *i2010 - Annual Information Society Report 2007*. [Online] Available at: http://ec.europa.eu/information_society/ eeurope/i2010/docs/annual_report/2007/sec_2007_395_en_documentdetravail_p.pdf [Accessed 26 February 2009].

Commission staff working document, 2008. Accompanying document to the communication from the Commission to the European Parliament, the Council, the European Economic and Social Committee and the Committee of the Regions. *Future networks and the internet - Indexing broadband performance* [Online] Available at: http://eur-lex.europa.eu/LexUriServ/LexUriServ.do?uri=SEC:2008:2507:FIN:EN:PDF [Accessed 26 February 2009]

DiMaggio, P. Hargittai, E. Celeste, C. & Shafer, S., 2004. Digital inequality: from unequal access to differentiated use. In: K. Neckerman, ed. *Social inequality*. New York: Russel Sage Foundation, pp. 355-400.

Dutton, W. H. Gillet, S. E. McKnight, L. W. & Peltu, M., 2004. Bridging broadband internet divides: reconfiguring access to enhance communicative power. *Journal of Information Technologies*, 19, pp. 28-38.

Ewing, S. & Thomas, J., 2008. Broadband and the 'creative internet': Australians as consumers and producers of cultural content online. *Observatorio (OBS*) Journal*, [Online] 2, pp. 187-208. Available at: http://www.obs.obercom.pt/index.php/obs/article/view/215/190 [Accessed 26 February 2009].

Haddon, L., 2000. Social exclusion and information and communication technologies: lessons from studies of single parents and the young elderly. *New Media and Society*, 2 (4), pp. 387-406.

Haddon, L., 2004. Cultural differences in communication: examining patterns of daily life. *Mobile Communication and Social Change*. Seoul, South Korea 17-18 October 2009 [Online] Available at: http://www.lse.ac.uk/collections/media@lse/whosWho/LeslieHaddon/KoreaPaperCulture.pdf [Accessed 26 February 2009].

Haddon, L. & Vincent, J., 2005. Making the most of the communications repertoire. Choosing between the mobile and fixed-line. In: K. Nyíri, ed. *A sense of place. The global and the local in mobile communication*. Vienna: Passagen Verlag, pp.231-40.

Hargittai, E., 2008. The digital reproduction of inequality. In: D. Grusky, ed. *Social stratification.* Boulder: Westview Press, pp. 936-944.

Lievrouw, L., 2002. Determination and contingency in new media development: diffusion of innovations and social shaping of technology perspectives. In: L. Lievrouw, ed. *Handbook of new media: social shaping and consequences of ICTs.* London: Sage, pp. 183-199.

Mason, S. M. & Hacker, K. L., 2003. Applying communication theory to digital divide research. *IT&Society*, 1 (5), pp. 40-55.

Menou, M. N., 2001. Digital and social equity? Opportunities and threats on the road to empowerment. *LIDA (Libraries in the Digital Age) 2001 Annual Course and Conference*. Dubrovnik, Croatia, 23-27 May 2001. Zagreb: Faculty of Humanities and Social Sciences, University of Zagreb.

Merton, R. K., 1973. *The sociology of science: theoretical and empirical investigations.* Chicago: University of Chicago Press.

Norris, P., 2001. *Digital divide: civic engagement, information poverty, and the Internet worldwide.* Cambridge, MA.: Cambridge University Press.

OECD, 2007 *Participative web and user-created content: web 2.0, wikis and social networking.* [Online] Available at: http://213.253.134.43/oecd/pdfs/browseit/9307031E.pdf [Accessed 26 February 2009].

OECD, 2008. *Broadband growth and policies in OECD countries.* [Online] Available at: www.oecd.org/dataoecd/32/57/40629067.pdf [Accessed 26 February 2009].

Parayil, G., 2005. Digital divide and increasing returns: contradictions of informational capitalism. *The Information Society*, 21 (1), pp. 41–51.

Tolbert, J. & Mossberger, K., 2006. *New inequality frontier: broadband internet access*. EPI Working Paper 275. Economic Policy Institute.

Thomas, F. Haddon, L. Gilligan, R. Heinzmann, P. & de Gournay, C., 2005. Cultural factors shaping the experience of ICTs: An exploratory review. In: L. Haddon, ed. *International collaborative research. Cross-cultural differences and cultures of research*. Brussels: COST, pp. 13-50.

Van Dijk, J., 1999. *The network society: social aspects of new media*. London: Sage.

Van Dijk, J., 2005. *The deepening divide: inequality in the information society*. Thousand Oaks: Sage.

Warschauer, M., 2003. *Technology and social inclusion: rethinking the digital divide*. Cambridge (MA): MIT Press.

Pedro Puga

Chapter seven. Governments go online: eGovernment and the digital divide

Introduction

The digital divide, also discussed in terms of social exclusion, has for some years been a topic attracting attention from scholars, while overcoming it remains a major challenge facing every government. Meanwhile moves towards developing eGovernment have been given a special attention as part of the i2010 initiative. To know the implications of one for the other may provide new insights into both topics and help find new ways to narrow the digital divide and to improve public services. Hence, this chapter will focus specifically on the significance for the digital divide of the fact that many governments are moving towards a relation with their citizens that is increasingly mediated by ICTs. It will:
– examine moves towards eGovernment, clarifying what different implications this has for different types of digital divide and illustrating this with examples from Portugal
– reflect upon the implications of the online services offered by governments for divisions within countries, and indicate what further research would be appropriate
– indicate what further research is needed to explore these in more depth.

First we look at some definitions of the digital divide and the way the debate about the concept is drawing attention to different types of social exclusion. Then the chapter looks at what eGovernment entails. What are the disadvantages and advantages of eGovernment? Moreover, does eGovernment actually lead to new inequalities in terms of access to public services? A number of scenarios are considered, which include reflections upon how cross-national differences may arise.

The chapter next tries to identify those who use eGovernment services and those who do not, using a secondary analysis of official statistics for the UK and Portugal to trace evidence of new forms of social division in this field.

Finally, we turn to the evolution of web 2.0, showing how governments are using this as a new arena for the development of eGovernment. It suggests new research for monitoring the development of web 2.0 eGovernment in relation to digital divide concerns.

Issues around the digital divide

The term 'digital divide' first appeared in the 90s and has become one of the most popular concepts used by scholars, consultants, companies and politicians. To start with an official body, the OECD defines the digital divide as:

> "The gap between individuals, households, businesses and geographic areas at different socio-economic levels with regard both to their opportunities to access information and communication technologies (ICTs) and to their use of the Internet for a wide variety of activities." (OECD, 2001, p.5).

However, in practice there are variations in the definition of the digital divide. For instance, Salinas (2003) refers it as being the gap between individuals and/or communities in terms of can or cannot use electronic information and communication tools, the internet being one of them, to improve the quality of life. Johnson Paul (2002) refers to the digital divide as the inequalities in access to digital infrastructures and services. Such a range of possible interpretations makes discussion of the concept problematic.

Several points can be drawn from this. First, one can talk about inequalities within societies (e.g. between individuals, households, communities in the sense of local or regional), or inequalities between societies (communities in a different sense). Both will be considered in this chapter.

Second, one can see two different emphases: (a) a focus on inequalities relating to ICTs more generally and (b) inequalities relating specifically to the internet. Some definitions, as above, and indeed some attempts to develop measures of the social exclusion combine both – e.g. Kastsinas and Moeck (2002), and Parker (2003) use personal computer ownership and internet access as indicators of the digital divide This chapter and the measures chosen follows those researchers who more specifically examine the internet digital divide.

Third, while some of the earliest debates where more orientated to 'access' (a term that can be observed in some of the above definitions) later commentators soon turned to other dimensions. For example, 6 and Jupp (2001) consider that placing access to hardware for internet use at the centre of the debate is a misconception of the digital divide, and argue that we should focus more on aspects like the real price of the equipment used to access the online world, as well as issues about content, skills, uses and social consequences. Once again, this chapter will examine these various dimensions beyond access.

While some of the earliest debates focused on internet users versus non-users (Katz & Aspden, 1998), it soon became clear that one can also examine inequalities between different internet users, for we can have 'divides' between those with different levels of skills, for example. This was already discussed when the majority of users only used narrowband in those early digital divide debates in the late 90s and early years of the 21st century. There are new potentially new divides between narrowband and broadband users, as noted in chapter 6 in this volume. Indeed, this reminds us that the digital divide is also a moving target as the internet itself develops, illustrated by the current web 2.0 developments.

eGovernment

If the previous sections provide a brief overview of relevant aspects for the digital divide debates, this one indicates the sections of eGovernment that are of interest. The European Community defines eGovernment as:

> "…the use of information and communication technologies in public administrations combined with organisational change and new skills in order to improve public services and democratic processes and strengthen support to public policies." (2003, p.7).

The concept of e-democracy, or electronic democracy, is assumed to be broader than that of eGovernment, referring to ICTs as tools to enable citizens to have greater access to information and services. Broadly speaking one can discern to major 'benefits' of these developments. One, of particular interest to political scientists, is that of political inclusion, implied especially in the emphasis upon 'democracy'. Using as term first used by Pippa Norris (2000), the position of the "cyber-optimists" on e-democracy relies on the assumption that ICTs facilitate a broad area of active citizenship, since they empower individuals by expanding their knowledge and ability to participate in broader processes of decision-making, including electronic voting.

However, in this chapter it is the second benefit that is of more interest, where governments offer services of various kinds in an electronic form. This can be illustrated with the case of Portugal. According to the UMIC/INE report on the Information Society (2007) in Portugal 60% of all Central Public Administration Bodies with internet access used it for communication with citizens. Of all Central Public Administration Bodies, 89% had a web presence, and among these 83% had their own website, 10% had a website integrated into the website of the Ministry and 1% had other types of website. Among them who did not have a presence on the internet, the most significant reason (26%) was a lack of employees with suitable abilities.

Providing information and allowing users to search for information are the main activities of eGovernment. UMIC/INE (2007) indicated that 98% of all Central Public Administration Bodies with a web presence provided institutional information about the Central Public Administration Body, 96% information about the services, 95% legislation and 62% access to databases. However, despite this major presence of public websites, only a small portion of individuals in Portugal go online to relate in some way to government – 42% of internet users go to the web to obtain information from public authorities' websites, suggesting that for the majority there is either some lack of awareness or motivation to do so. In other words, even amongst those who do have internet access, who are advantaged in the traditional digital divide, a new citizenship divide appears with the emergence of the e-citizen.

Face-to-face vs. online: advantages and disadvantages

So far this chapter has indicated the areas of interest. Now we turn to see what the eGovernment debates might learn from the digital divide literature, and illustrate what this means with examples from everyday life.

Governments tend to overstress the simple benefits of eGovernment initiatives, when it really is a more complex picture. For one thing, the problem is that optimistic visions fail to take into account sufficiently the issues raised in the social exclusion literature. People without internet access are at a disadvantage compared to those who have access. Even for those with internet access one can foresee inequalities that may arise from difference in terms of online skills, which might have real effects on their ability to use government provided public services. And to build upon the point made earlier, those who have narrowband access are potentially at a disadvantage compared to those who have broadband access. Although a limited number of government services require broadband, this may increase in the future.

Some people who may be disadvantaged in their access to eGovernment may not be disadvantaged more generally. However, as has been pointed out by other researchers, inequalities in terms of different citizens' relation with government services offered through the internet are mainly the expression of the previous social and economic inequalities:

> "The Digital Divide is not digital; it's the social and economic divide which is reinforced by technologies that exacerbate the potential to exclude people." (Kate Oakley, cited in Craig and Greenhill, 2005, p.19).

To sum up so far, eGovernment is a potentially new field in which the inequalities might be experienced, where adding more services might (further) disadvantage those who are already deprived of material goods or skills. These potentially new types of inequalities are still not well documented. Although the relation between the digital divide and eGovernment is not a total novelty in the academic field, few researchers have addressed this relation and both more qualitative or quantitative data are needed in order to show in more detail the forms and extent of these inequalities that we know in principle exist.

Before progressing in this argument, it is important to add some qualifications that question whether eGovernment is always and totally to be embraced. There is a range of advantages offered by having services online. For instance, governments often set different deadlines for those filling in tax forms online vs. on paper: in the case of UK and Portugal, governments give more time to those who want to submit their taxes online, and in doing so provide an advantage for the online users. Usually, those who have filled their taxes online have a response from the government sooner than those who do it by paper. In other words, access to eGovernment services can provide some with benefits, but they can also mean a relative disadvantage for others not using them.

For instance, when one accesses government services online one can do it more quickly than having to go travel to visit the offices of a government agency. However, one then has to rely on the information that the website provides. Some government websites have an HELP menu or FAQ[59] menu, but these do not always answer all potential questions – only the pre-designed questions have answers, which means that the sites have limitations. In contrast, someone going to visit a public agency can explain in his or her own words any problems or ask for clarifications – which might not be addressed by the online FAQs.

To take a second illustration, when filling out a paper form it is possible to stop whenever one wants, but that is not always the case with online forms. Sometimes it is possible to save what has been done but that is not always the case – some software only allows one to save a part of the form. Sometimes the design of the software assumes that the time one needs to fill in the form is equal for everyone and some software allows only a limited time for the form to be filled out – often one needs to start all over again if it takes too long. The principle shown by both examples is that traditional interactions with administration (be that through face-to-face contact or filling out paper forms) often provide more flexibility than the online version.

Basically, different people have different skills but websites treat people as equal, giving them the same time and conditions for filling forms, or to access some services. For instance, among the Portuguese Central Public Administration Bodies websites, only 48% of have accessibility for citizens with special needs.

So there are advantages and disadvantages for both offline and online forms of administration. Moreover, these illustrations show the fact that the offering of services online is not just creating potential disadvantages for non-users of the

---

59 Among the Portuguese Central Public Administration Bodies with web presence, 54% have user support (FAQs, Helpdesk, etc.).

internet but also in some cases creating problems for internet users, when we compare both online and offline versions in more detail.

Reducing social exclusion? Scenarios and research questions

Having established and illustrated certain principles in this area, the question is: what effects will eGovernment have upon social and political exclusion in practice? One can start to address this by considering a number of different scenarios and asking what the relevant evidence might look like.

The awareness by individuals that ICTs like the internet may facilitate the relationship to their government can in principle act as a catalyst for individuals to make the effort to access the internet, or to learn how to use it. In both these cases, governments need to, and some do, create possibilities for the individuals to have internet access kits and training in the use of the internet. This might narrow the digital divide and in doing so may reduce the social and political exclusion of those with special needs - for instance, older people with mobility problems, isolated or almost isolated rural inhabitants, etc. In this scenario, the offer of eGovernment services can actually help to alleviate the digital divide.

To take a particular example, in Portugal there is a governmental programme called 'Plano Tecnológico' ('Technological Plan'), part of which involves the implementation of online public services. One of the most well-known parts of this programme is the project 'Magalhães', giving students in the first grade at school the possibility to purchase mobile computers at very low prices. Another project gives teachers the option to purchase mobile computers with internet access at low prices. Yet another project is the 'Iniciativa Novas Oportunidades' ('New Opportunities Initiative'), which enables youngsters and adults who left school at an earlier stage to access training programmes and gain more qualifications. For those who choose to take up this offer, they can also acquire mobile computers with internet access at low prices.

In evaluating such initiatives one needs to know whether these are indeed having the effect of persuading people to gain more qualifications and purchase computers and internet access. And if so, to what extent? Are these initiatives successful in terms of creating an awareness of the importance of being able to relate to the government online? Moreover, if one considers proxy-users who use eGovernment (i.e. people who already have and can use the internet acting on behalf of non-users), do such proxy-users get to know more about or become more interested in going online through that experience of proxy-use? Will they move from being proxy users to directly accessing eGovernment services?

Of interest for cross-cultural analysis, the way that various national governments promote the use of the internet might well differ and have different effects on the relationships between citizens and governments. Countries have specific social and cultural characteristics, and their citizens might relate differently to online public services. Governments need to map out these features of their population, identifying the best way to offer public services suited to that population as well as the best ways to persuade people about the benefits of going online and especially about the benefits of relating to government through the web (as long as there are advantages compared to face-to-face contact).

For example, consider the implications if the Government in one country strongly subsidises the use of the internet for current non-users, but does not give citizens the right information about the existence and benefits of eGovernment. Whilst this might possibly be narrowing the divide between users and non-users, it might not narrow any divide between those who use public online services and those who only access to face-to-face services. In another country, where eGovernment is more strongly promoted, that second level of the divide may narrow.

In addition, different countries might have different social inequalities in terms of access to internet, skills, and awareness of eGovernment, and so there is the need to know the details of these inequalities in terms of age, gender, income, education, etc. The way governments may scope the problem may also, of course, led to different consequences, narrowing or not narrowing the digital divide among the various fractions of their population. If eGovernment promotion in one particular country focuses mainly on youngsters and forgets elder people in terms of ICT promotion, the Government may be broadening the digital exclusion of elder people and narrowing the divide for youngsters. In other words, difference in the promotion and development of eGovernment services, if the focus is on giving advantages to some segments of society and disregarding others, may lead not only to divides within countries but also differences between countries.

In a second scenario, the citizen who cannot fill out forms electronically because he or she does not have an internet connection, does not have the economic conditions to do so, does not know how to use the eGovernment service, or just prefers to achieve these goals face to face may not feel socially excluded, for he or she can go to the 'real' government agency and fill out this form there. Thus such people may still be in control of their own lives and may be able to fully participate in the society. Of course, they are still set apart a technological service that might make life easier (for instance, filling out form in the comfort of their own home) and provide quicker services. Moreover, to put this in perspective, even if one service is only accessible by the internet, such a citizen is not in general socially excluded, but rather excluded by the government from particular administrative services. They may well be active citizens in other respects, engaged in community social life and even play the role of political actors. In this scenario, there may be a disadvantage but one would hesitate to say that this is enhancing the digital divide.

In this scenario, any research would have somehow to evaluate the degree to which non-use of eGovernment options creates any disadvantage. Ultimately, is it a barrier to some people using some government services altogether? How important is that overall in their life? Does this vary depending on the circumstances of different segments of the population? For example, it may be a greater disadvantage for those who live in rural areas, for they need to travel every time they need to go to public services, and so, for instance, they might be less active in terms of citizenship. But, such forms of digital exclusion may not be so great a disadvantage for an individual who lives in an urban area with easy access to public services.

Once again, there is a cross-cultural element – the last example being relevant because different countries have larger and small rural populations, living at different distances from urban centres. In principle, one can imagine, in this scenario, that lack of access to some eGovernment services in one country may be experienced as more or less of a problem compared to lack of access in another, for whatever reasons, be they geographical factors as noted above, or cultural.

In a third scenario, if a particular only-online service is very important for one's social life or even for active citizenship, that is a different matter. Moreover, this problem can grow as if more and more services and goods will only be accessible through the internet. In this case, the negative aspects may increase as some fractions of society may be entirely excluded from these public services. For instance, if the government offers a website to help job-seekers but does not provide this information face-to-face, then this favours the unemployed with internet access (e.g. more information, easy access to information, more possibilities in applying for a job) when competing for those jobs.

Once again, for the cross-cultural dimension one needs to understand the way that the different countries are developing eGovernment: as a complementary service to the face to face one (i.e. as another option, another route to government services) or as unique services that will displace the current face-to-face options. In addition, even in countries where complementary online services are offered one would have to research whether there are nevertheless country variations in the difference between the online and offline versions. For instance, we noted earlier that in some countries like Portugal, tax forms can be filled out and submitted online for a much greater period of time than in is the case for paper tax forms. But would this time difference between the online and paper versions be the same for all countries? If it varies it can make the 'advantage' on online services lesser or greater, depending on the country. This is another example how differences in the implementation of eGovernment might lead to different digital divides in different countries.

Evidence of digital divides

Despite the available data we saw earlier showing what public services governments are putting online, and even some data on the use of public services, it is still generally difficult to find measures of the relationship between the availability of public services online (or whether it is necessary to use online public services) and whether this is increasing the number of new users or leading to new skills. That said, there is evidence to illustrate the nature and extent of different divides.

First, Table 1 shows the overall differences between the UK and Portugal for this as regards the use of three different eGovernment services, where the only significant difference, but a substantial one that underlines country differences, is that the Portuguese seek less information from websites.

	Downloading official forms	Obtaining information from public authorities web sites	Sending filled in forms
European Union (27 countries)	16	26	12
Portugal	12	15	13
United Kingdom	14	26	12

Table 1: Percentage of individuals who used the internet in the last three months for..., 2008.
Base: All individuals age 16-74
Source: Eurostat - Computers and the internet in households and enterprises

Tables 2-4 show the breakdowns by age and gender, so that we can check how much the demographic patterns are similar across countries. In the EU27 generally, those aged between 25 and 64 are more likely to have used the Internet in the last three months for downloading official forms, obtaining information from public authorities' websites and for sending filled in forms.

The most noticeable difference, running counter to perceptions of elderly people's disinterest in the internet, is that in Portugal the 65-74 year olds are only a few percentage points behind those core younger groups. (NB there is a lack of data for individuals in the UK aged 65 to 74[60]). Hence, there is some

---

60 There are some missing or unreliable data (referenced as ":") in the Eurostat statistics, so the analysis of Eurostat data can only take into account the existing data.

difference in age-based use between countries. For both countries, males are slightly more inclined to use all three services.

	Age					
	16 to 24	25 to 34	35 to 44	45 to 54	55 to 64	65 to 74
European Union (27 countries)	17	31	29	28	28	20
Portugal	13	38	35	32	33	35
United Kingdom	16	19	20	20	19	:
	Gender					
	Females, 16 to 74 years old			Males, 16 to 74 years old		
European Union (27 countries)	25			28		
Portugal	27			30		
United Kingdom	16			21		

Table 2: Percentage of individuals who used the internet in the last three months for downloading official forms, 2008.
Base: Individuals aged 16-74 years old who used the internet in the last three months
Source: Eurostat - Computers and the internet in households and enterprises

Although shortage of space means the equivalent tables for other socio-demographic variables are not shown here, there are certain similarities between the UK and Portugal in terms of general trends, even if the percentage differences between the countries vary a little. It is clear that those individuals with upper incomes are more likely to have used the internet for downloading official forms, obtaining information from public authorities' web sites or for sending filled in forms, in both Portugal and the UK. When it comes to education, the higher the level of education the more one is likely to have used the internet to achieve these same three goals – in both countries. And the same pattern holds for non-manual workers, who are more likely to use all three services, and the category of employees, self-employed and family workers (compared to the unemployed, students and retired).

	Age					
	16 to 24	25 to 34	35 to 44	45 to 54	55 to 64	65 to 74
European Union (27 countries)	29	47	46	45	45	34
Portugal	19	45	44	40	45	34
United Kingdom	28	34	38	34	39	28
	Gender					
	Females, 16 to 74 years old			Males, 16 to 74 years old		
European Union (27 countries)	40			43		
Portugal	34			38		
United Kingdom	32			36		

Table 3: Percentage of individuals who used the internet in the last three months for obtaining information from public authorities' web sites, 2008
Base: Individuals aged 16-74 years old who used the internet in the last three months

	Age					
	16 to 24	25 to 34	35 to 44	45 to 54	55 to 64	65 to 74
European Union (27 countries)	12	22	21	21	21	17
Portugal	12	39	38	38	39	34
United Kingdom	14	15	16	15	19	:
	Gender					
	Females, 16 to 74 years old			Males, 16 to 74 years old		
European Union (27 countries)	18			20		
Portugal	29			33		
United Kingdom	15			17		

Table 4: Percentage of individuals who used the internet in the last three months for sending filled in forms, 2008.
Base: Individuals aged 16-74 years old who used the internet in the last three months
Source for Tables 3 and 4: Eurostat - Computers and the internet in households and enterprises

The picture is more varied when considering the type of areas one lives in. In both countries individuals living in sparsely populated areas (less than 100 inhabitants/km²) have smaller rates of all three activities. However, the patterns vary more between the countries when comparing rural, intermediate and urban areas.

Two other points pertinent to the general argument of this chapter are worth mentioning from these tables. First, Portugal and the EU 27 show higher rates among those with broadband access than those with non-broadband access for downloading official forms, obtaining information from public authorities' web sites or for sending filled in forms, although we have no data for the UK. Hence, broadband access, not just general internet access, is making a difference to eGovernment use. Second, the fact that ICT professionals are amongst the ones that use eGovernment services shows that ICT skills also matter, and that access to eGovernment creates inequalities in the "power" to reach government services.

To summarise this section, in general we can trace a common 'profile' of those likely to be an eGovernment users: they are more likely to be male (only slightly more likely in this case), aged between 25 and 64, live in urban areas, have higher education, be non-manual workers, have high incomes, and be ICT professional. These figures also underline the divides to be considered when it comes to eGovernment usage. In this sense, eGovernment can be thought as one more service where different segments of society are set apart.

Future eGovernment digital divides: eGovernment 2.0?

Finally, the chapter ends by looking beyond the eGovernment services that have been established for some years to consider newer developments online. Specifically, as a case study, it examines how governments are now looking at virtual worlds as another form of online representation.

In Portugal, Antonio Costa, current Lisbon City Mayor, in his campaign for the City Hall, created an avatar in the virtual world Second Life (SL) (an avatar more lean and without glasses, but despite that, looking 'exactly' like him) in his online campaign headquarters[61]. In France, the four main candidates for the presidential elections also took part in this virtual world with their respective campaign headquarters and avatars[62]. In the USA, Congress created a virtual

---

61 Zamith, F., 2007. António Costa em campanha no second life, *Diário de Notícias* [internet] 7 July. Available at: http://dn.sapo.pt/2007/07/07/nacional/antonio_costa_campanha_second_life.html [Accessed 26 January 2009].

62 Moore, M., 2007. French politics in 3-D on fantasy web site, *The Washington Post Online*, [internet] 30 March. Available at: http://www.washingtonpost.com/wpdyn/content/article/ 2007/03/29/AR2007032902540.html [Accessed 26 January 2009].

version of the U.S. Capitol[63]. The project opened with the release of a video streaming with an online presence of Rep. George Miller, of California, chairman of the Democratic Policy Committee. The government of the Maldives Islands was the first one to open Virtual Embassies in Second Life, in the so-called 'Diplomacy Island'[64]. Sweden built a virtual embassy, the 'Second House of Sweden'[65]. Recently, Estonia also opened an embassy on Second Life[66]. Other government agencies and public institutions are now represented in Second Life, such as Ontario Public Service[67]. Governments, parties and politicians, are clearly looking into new ways to reach citizens.

As the internet evolves so to does eGovernment and that too can be creating new digital divides, not only between those who use and those who do not use internet, or now between narrowband users and broadband users (as was indicated in earlier tables). The presence in virtual worlds exemplifies what might be a next turning point for eGovernment.

Web 2.0 challenges users and governments, as a door to new worlds yet to be discovered. If these worlds "have" people, they can be a platform for public services. In fact, existing portals and websites can actually seem "static" compared with some new web 2.0 features. This may be an incentive for governments to engage with these new developments: it is not only Second Life that is gaining importance in the field of eGovernment, but also wikis, blogs, social networks, etc., as noted by some analysts:

> "Actions such as jumping on the wiki bandwagon, using technologies such as Ajax for richer user interfaces or diving into virtual worlds to entice the so-called "digital natives" will result in a sudden awakening for governments. We expect several governments in developed economies to establish virtual government strategies that define how to participate in a variety of virtual communities, ranging from internal ones that engage employees, to external ones where they will reach out to constituents.

---

63 Business communicators of Second Life, 2007. *United States Congress enters Second Life. 100:00 Hours and counting.*
64 The Times, 2007. Tiny island nation opens the first real embassy in virtual world, *Times Online* [internet] 24 May. Available at: http://technology.timesonline.co.uk/tol/news/tech_and_web/article1832158.ece [Accessed 26 January 2009].
65 Sweden.se, 2007. *Sweden opens virtual embassy 3D-style* [Online] (Updated 30 May 2007) Available at: http://www.sweden.se/eng/Home/Lifestyle/Reading/Second-Life/ [Accessed 26 January 2009].
66 Estonian Ministry of Foreign Affairs, 2007. *Estonia opens embassy in virtual world Second Life* [Online] (Updated 4 December 2007) Available at: http://www.vm.ee/?q=en/node/643 [Accessed 26 January 2009].
67 Wiki Second Life, 2009 *Real Life Government/Examples*. [Online]. (Updated 22 May 2009) Available at: http://wiki.secondlife.com/wiki/Real_Life_Government/Examples [Accessed 15 December 2009].

Such efforts will provide value only if they are very well-focused and conducted, at least initially, in the context of gated communities where governments can exercise some degree of control." (Gartner, cited in MacManus, 2007)[68].

Linden Lab, the company that runs Second Life, has a special website for helping enterprises, governments and educators to conduct activities and own places on SL. In the government section, it says:

> "Many government agencies and departments, including military organizations, at all levels around the world use virtual worlds on the Second Life Grid™ for a variety of programs and activities.
>
> For instance, the National Oceanic and Atmospheric Administration (NOAA) and the Earth System Research Laboratory (ESRL) have a facility that provides interactive educational demonstrations about the ocean and weather. Visitors can ride submarines, experience a tsunami, or check out real-time weather maps." (Second Life Grid, n.d)[69].

The development of eGovernment means not only that governments are creating new services online and improving existing ones, but also they are looking for something new on the web, new possibilities to expand their presence there. In doing this, governments are creating services that require skills in using the internet, better hardware and software and faster connections to that internet, all of which have the potential to create more inequalities in access to eGovernment services, to disadvantage those who cannot fulfil the needs required for accessing them.

Conclusions

Accessibility to government services through the web may be an advantage for internet users, for they can access these public services at distance. But eGovernment is not only about new advantages but also about new disadvantages. eGovernment provides a new dimension to the digital divide problematic, for it adds yet new divides between those who access public services online and those who do not. This was made clear by the statistics provided by Eurostat showing divides by age and gender, but also clear divides by income, education level, urbanisation level, etc. The development of

---

68 Gartner cited in MacManus, R. 2007. *EGovernment meets web 2.0: goodbye portals, hello web services* [Online] (Update 5 November 2007) Available at: http://www.readwriteweb.com/archives/e-government_meets_web_20.php [Accessed 26 January 2009].
Second Life Grid, n.d. Government. [Online] Available at: http://secondlifegrid.net/slfe/government-use-virtual-world. [Accessed 26 January 2009].

eGovernment, it is often suggested, is closely linked to the need to narrow the digital divide in order to diminish inequalities, yet figures show how in certain ways it can widen them.

This proves, despite the need for more data as mentioned earlier, that there are segments of countries' populations that remain apart from eGovernment. Hence debates about new eGovernment initiatives should pay attention not only to the characteristics of internet non-users but also to internet users who are eGovernment non-users.

In this respect differences between countries have also to be taken into account, especially when the EU itself is promoting eGovernment's development across member states. The implementation of eGovernment can differ across countries, which can have implications for the different digital divides scenarios outlined in this chapter.

Empirically, there is a need for more evidence in terms of more statistical data, clearly identifying attitudes to and representations of eGovernment – both between those who do and do not use the internet and also within the user population. Since this chapter has noted the pros and cons of eGovernment itself, these studies needed to be complemented by ones trying to assess whether and for whom the non-use of eGovernment services actually constitutes a disadvantage, a form of social exclusion and if it does, how significant is this within the overall lives of those involved.

To assess eGovernment, longitudinal studies, or at least repeated studies, focusing on the impact of initiatives taken by governments across time are necessary in order to understand how different initiatives, and different ways of promoting and implementing eGovernment, may affect the usage of the internet for eGovernment purposes among existing internet users, and in so doing may affect the evolution of digital divides.

The final argument is that further research on this topic should start to examine the evolution of the relation specifically between eGovernment and web 2.0 developments. While the intention of governments is often to explore new digital channels for increasing citizen involvement and engagement, for overcoming political and social exclusion, we must be vigilant in asking in what ways this may be a double-edged sword with the potential for new forms of inequality.

Ultimately understanding the relationship between eGovernment developments and the digital divide debates is crucial as governments try to promote digital inclusion, at the same time as they themselves becoming more digitally present in the online world.

Bibliography

Craig, J. & Greenhill, B., 2005. *Beyond digital divides? – The future for ICT in rural areas.* [e-book] Cheltenham: CRC. Available at: http://www.demos.co.uk/files/ruralbroadband.pdf [Accessed 1 November 2008].

European Commission, 2003. *"The role of eGovernment for Europe's future" Communication from the Commission to the Council, the European Parliament, the European Economic and Social Committee and the Committee of the Regions*, Brussels: European Commission. Available at: http://ec.europa.eu/information_society/eeurope/2005/doc/all_about/egov_communication_en.pdf [Accessed 1 November 2008].

Kastsinas, S.G. & Moeck, P., 2002. The digital divide and rural community colleges: problems and prospects. *Community College Journal of Research and Practice*, 26, pp.207-224.

Katz, J. & Aspden, P., 1998. Internet dropouts in the USA. The invisible group. *Telecommunications Policy*, 24(4/5), pp.327-339.

Norris, P., 2000. Democratic divide? The impact of the internet on parliaments worldwide. *Annual Meeting of the American Political Science Association.* Washington DC, 31 August-3rd 2000.

OECD, 2001. *Understanding the digital divide*, [pdf] OECD Publishing. Available at: http://www.oecd.org/dataoecd/38/57/1888451.pdf [Accessed 1 November 2008].

Parker, B., 2003. Maori access to information technology. *The Electronic Library*, 21 (5), pp.456-460.

Paul, J., 2002. Narrowing the digital divide: initiatives undertaken by the Association of South-East Asian Nations (ASEAN). *Electronic Library and Information Systems*, 36 (1), pp.13-22.

Perri, Jupp Ben, 2001. *Divided by information? The "digital divide" and the implications of the new meritocracy.* London: Demos.

Salinas, B. Romelia, 2003. Addressing the digital divide through collection development. *Collection Building*, 22 (3), pp.131-136.

UMIC – Knowledge Society Agency/INE - National Statistics Institute, 2008. *The information society in Portugal 2007.* [Online] (18 December 2007) Available at: http://www.umic.pt/index.php?option=com_content&task=view&id=3034&Itemid=408 [Accessed 1 November 2008].

Jorge Vieira

Chapter eight. Differences in the experience of music 2.0

Introduction

> "After coming to the London School of Economics for my study visit I worried about bringing with me "some" music. I looked at my iTunes. Roughly 16GB of music. This is equivalent to 40 days of non-stop music playing in a small and portable laptop or digital player. Is that enough? Rather good, but you can never have too much music, because you might want to hear that specific music... But I also know that if I am in the mood for a particular listening experience that I do not currently have on my hard disk, I can, almost certainly access it easily and quickly online." (Jorge Vieira)

This kind of thinking and practice is the essence of the music 2.0 concept, implying abundance instead of scarcity, ease of access, a culture of music taste sharing bound with a sense of fluidity and mobility.

Given these contemporary developments, the first question concerns if and how the increase in the digitisation of music and the enhancement of network communication on a global scale are together changing people's relationship with music. Our first hypothesis is that they are. And the path of change is leading towards a greater degree of multiplicity and complementarity of practices, between multiple offline and online supports (sometimes termed 'crossmedia' consumption – see chapter two). This chapter examines the contours of this new experience using Portuguese data, showing the variety of levels on which users are engaging with the world of music 2.0, or not, where multiple forms of music consumption are distributed very unevenly across the entire population.

Normally one does not associate music with discussions of digital divides (as outlined in chapters six and seven especially). These early social exclusion debates examined unequal access to resources, observing how some may be disadvantaged relative to others, especially in terms of opportunities in life. Certainly in earlier formulations, these discussions did not usually focus on, say, the distribution of resources from what have been called the 'cultural industries'. Arguably they should have. Certainly media studies in general have sought to show how important a role the media play in our lives, and so even if one hesitates to talk about 'divides' given diverse media tastes and interests, the existence and implications of 'uneven' (Haddon, 2004) access to media, as a source of culture and leisure, merits attention. But, of special interest here, is that those digital divide studies provided a set of measures – of access, use, skills – that we can draw upon when trying to chart the varied forms of music 2.0 experiences.

First let us think about the particularities, and significance, of music more generally, to put music 2.0 into context. Music is one specific instance of media content. It is important to stress the social relevance of music, because 'popular music is a primary if not the primary, leisure recourse in late modern society' (Bennett, 2001, p.1). For many, music matters. It is, for most of us, omnipresent, whether you want it or even realise it. Music is everywhere and reaches a wide range of individuals. We are exposed to music in a more or less direct or indirect way, from the music in the underground station to the music selected by our chosen radio station. Moreover, numerous writers (e.g. Frith, 1996, Shuker, 2008; Longhurst, 2007) have pointed out that music is an important dimension of the processes that (re)configure social identity. Simon Frith states that:

> "music constructs our sense of identity through the direct experiences it offers of the body, time and sociability, experiences which enable us to place ourselves in imaginative cultural narratives." (Frith, 1996, p.124).

We can observe these dimensions in many studies conducted on those 'sub-cultures' structured in terms of communities of musical tastes – from hip-hoppers, ravers, emos to heavy metal sub-cultures (for example, Gelder, 2007; Hodkinson & Deicke, 2007) – as well as in the way that music as an audio-environment works has an important factor in everyday social interactions (DeNora, 2002), from the track one chooses at home for coffee with friends to the soundtrack one comments on in the shopping mall.

But there is a second reason to consider music as being a strategic case study. Much of the earlier discussion on the digital divide focused on the first decade since the internet became a mass market. But now, researchers are drawing attention to the new opportunities opened up as the internet has evolved, both in terms of infrastructure through broadband and in terms of applications through the so-called web 2.0 applications – which includes some of the sharpest developments around music. However, as such new developments occur, this inevitably means that some people have more options than others, some have more resources than others, some have more skills to take advantages of new possibilities and develop new forms of consumption and production, as in current discussions of prosumers[70] (Toffler, 1981). Whether or not one goes as far as to characterise these as new forms of digital inequality, some have at least referred to a 'participation gap' (Jenkins, 2006) while others associated with the social exclusion literature have indeed used the words 'participation divide' (Hargittai & Walejko, 2008).

---

70 A contraction of the words "Producer" with "Consumer".

One hypothesis is that the new forms of social appropriation and domestication of the dematerialised consumption associated with music 2.0 will take place to a greater degree among the younger generations of 'digital natives' (Prensky, 2005) – the social agents that grew up with and are currently immersed in a whole system of digital media, linked into online networks, and characterised by social representations, media diets and technological devices radically wider and distinct from the "radio generation or the 'television generation'. If such youth are at the forefront of such developments it is worth adding even at this early stage that it is within this subgroup that we may well also find the sharpest perception of uneven access and differentiated experience in terms of music consumption, since it is among this group that the percentage of individuals using music 2.0 is much higher.

Before going on to examine the data on consumption practices it is important to put music 2.0 into some historical perspective to appreciate the extent to which it is new but also the extent to which is also located within a continuum of previous innovations.

Technological innovation, music industry and music consumption: the path to music 2.0.

While music 2.0 may be a new concept, it actually reflects a much longer, process of evolution within the music industry that has gradually changed music consumption practices. Here we note some of the key moments in the field of reproduced music more generally in order to indicate specific ways in which the experience of music has changed over time.

Until the invention of the phonograph, by Thomas Edison and Emile Berliner, and the patenting by Emile Berliner of the Gramophone, both circa 1887, musical consumption had to be in the form of live music. Presence was *a sine qua non* condition of listening to the music practices of others, making it an experience that was ephemeral, unique and unrepeatable. The advent of recorded music had radical implications for the field of music, establishing a shift from the rarity of the occurrence (i.e. live music) to the possibility of technical reproducibility (Benjamin, 2008) (i.e. recorded music). This step allowed the reproduction *ad infinitum* of a given past musical artefact and removed the strings that tied live music to a particular moment of consumption, changing profoundly how we now hear and relate to music. In addition to the blurring of time and space constraints, these inventions led to the opening of the once public musical enjoyment to a private sphere, allowing an individualized hearing. Music, until then available only to an elite that could pay musicians, was transformed from a service to a cultural product – a commodity.

The introduction of radio in the 1930s opened up such listening practices to a larger number of citizens, taking one more step towards the mass consumption of music. The advent of the audiotape by Philips in 1963 was the next major step leading to the spread of the home audio copy culture. This innovation introduced deep changes in music distribution, allowing a more active form of consumption (for instance in terms of editing/recording practices) and the individual-to-individual mediation (i.e. the mixtape sharing culture), promoting the creation of communities structured in terms of specific, socially shared and cultivated tastes - with clear implications in the culture sphere, for social identities and lifestyles.

The next set of milestones in people's relationship with music was set with the portable cassette player, the ordinary Walkman, commercially released by Sony in 1979 (Du Gay, et al., 1997). This technology revolutionised the music field in the sense for eating a different relationship with the music, linked to physical mobility, introducing the concept of 'music on the go'. This was followed by the launch of the Compact Disc, which allowed faster than real time precise and perfect digital copies, providing the subsequent spread of music in a dematerialised format (from MP3 to FLAC, to name but a few) and allowing an even greater degree of mobility in terms of small, light portable music players like the iPod (Levy, 2007). These digitising algorithms expanded the possibilities for sharing among networks, facilitating online exchange of small compressed files.

But, one could ask, are the present sharing practices totally different from those of preceding times? Are there any real specificities? Is there any uniqueness to these digital practices? We were already sharing music in the past. We were already listening to music in transit. Arguably in the music 2.0 era all of these practices are taken to yet another level – especially in terms of accessibility.

Many authors defend that we are witnessing a shift in a music paradigm, from tangible good-like commodity to an intangible digital service-like one (Leonhard, 2008; Styvén, 2007).

Digital music consumption in Portugal

Since music 2.0 is relatively new, there is little internationally available data on its take-up. However, one Portuguese survey asked some relevant questions. The following Portuguese data were gathered from the national questionnaire 'Network Society' 2008 ObcrCom – Portuguese Observatory for the Media. The survey, involving face-to-face interviews, covered to a representative sample of the population living in continental Portugal, aged 15 years of more. The sample universe was the whole population, as mapped by the results of General Census of Population – Census 2001 – conducted by INE. Individuals were selected by setting quotas from a combination gender, age, education, region and residence/size of population clusters. Starting from an initial matrix of region and habitat, a large number of sampling points were selected at random, where the interviews were conducted according to the above quotas. The final sample consisted of 1,039 interviews. The fieldwork was conducted in June 2008 and implemented by GfK Metris Portugal.

Daily time spent listening to music

In order to understand the importance given to music in daily life, and put it into perspective, figure 1 shows the daily average listening times in Portugal. Clearly for the majority (51%) music is only a part of their everyday life and for a sizeable minority music starts to play a larger role, with 30% stating that they listen to music more than one hour per day.

On the other hand, we can observe that 19% of the population do not listen to any music.

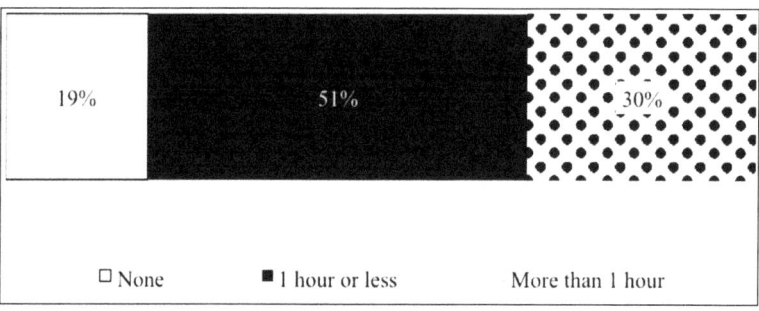

Figure 1: Daily average time spent listening to music. Portuguese population above 15 years Source: Network Society in Portugal, 2008. OberCom Base: 920

Crossing this variable with age groups we can observe that older groups attribute less importance to music in their everyday life – for instance, 40% of those 65 or older do not listen actively to music. On the other hand, one can see a more frequent role of music in the everyday life of the younger age groups. Here the proportion of individuals detached from music is smaller and time spent listening to music is greater (Figure 2), a trend similar to the findings of previous studies (Shuker, 2008).

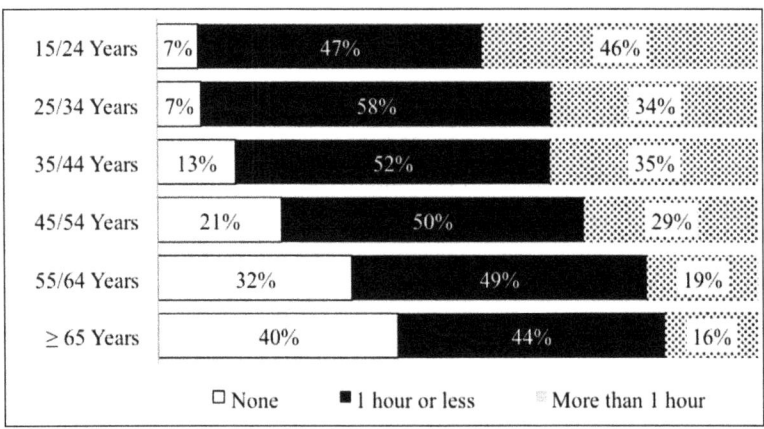

Figure2: Daily average time spent in listening to music by age group.
Source: Network Society in Portugal, 2008. OberCom
Base: 920

Internet penetration rate and access issues

When asking more specifically about the take up of music 2.0, one obvious starting point is whether one is an internet user at all, since the world of music 2.0 is unavailable in its fullest form for non-users.

In Figure 3 we can see that more than half of the Portuguese sample had never used the internet at all. But arguably we can add to this figure the 'rare' users: individuals that have been online, but only in an indirect and mediated form with the help of others (3%) and those that have gone online only once for experimental purposes (4%). If we now analyse in more depth those who say that they are online, we can also note that 4% of these are really partial users, divided into the 1% that use only the e-mail functions and 3% who use only the internet without having an e-mail account (which is a relevant requirement for sharing practices). Therefore we are dealing with 'full' internet use by around 37% and 40% of the population, meaning that for nearly two-thirds of the Portuguese population the world of music 2.0 is not directly accessible to them.

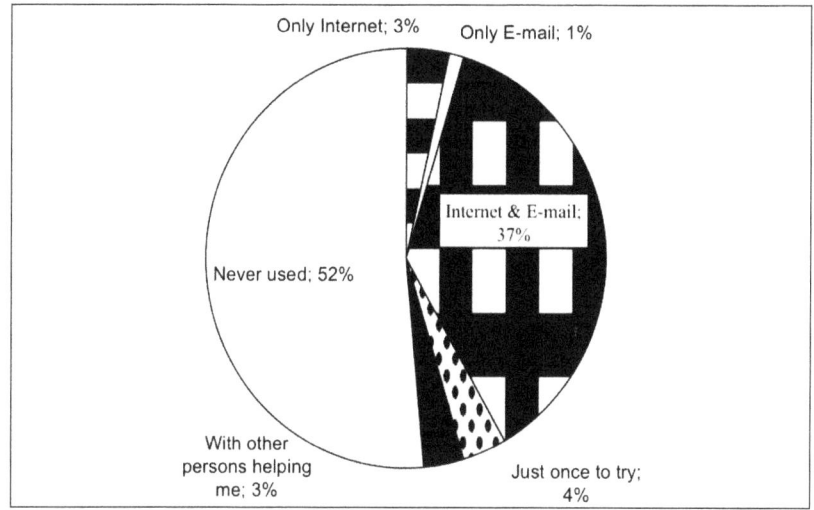

Figure 3: Have you ever used internet? Portuguese population above 15 years.
Source: Network Society in Portugal, 2008. OberCom
Base: 1039

However, the access picture is even more complex. For example, some digital divide studies referred to the proxy use of the internet (Haddon, 2004). We can follow up on this idea here because it is possible to benefit from the existence of music 2.0 without actually downloading oneself. Someone can 'rip off' (i.e. copy) digitally a CD collection to fill his MP3 player or else acquire MP3 files from the collections of friends or family.

So when considering music 2.0 we can also include these types of proxy users who are, maybe to a lesser degree, exploiting the potential of music 2.0.

Hardware. MP3 player penetration rate

Beer (2008) claims that MP3 players are reconfiguring music, first by making it available in a dematerialised form, with further implications for archiving and collecting music libraries, and second, in terms of music's omnipresence/ubiquity, both in time and space and public and private domains.

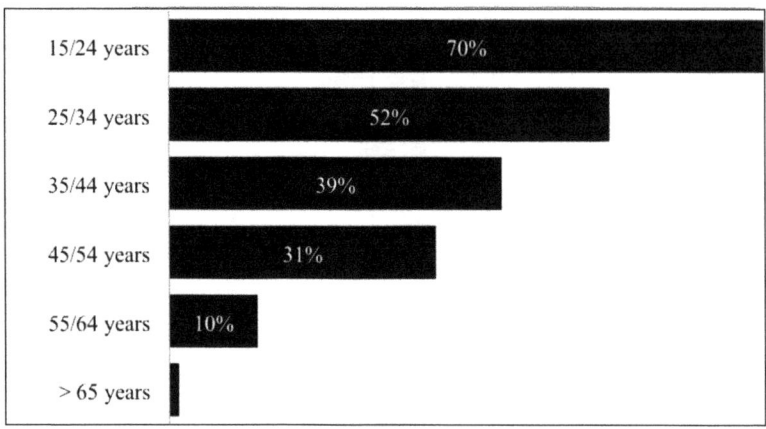

Figure 4: MP3 player ownership by age groups.
Source: Network Society in Portugal, 2008. OberCom Base: 1026

To get some idea of the extent of the spread of this technical apparatus in Portugal, 35% of the population have an MP3 player. This is a fairly good penetration rate and indicates a fast growth rate if we take in account the fact that in 2006 the portion was only around 12% of the population. And it is a remarkable diffusion given the novelty of this new personal stereo system, even if we acknowledge the fall in price levels.

If we now examine these data in relation to age, Figure 4 shows how the penetration rate grows as we descend through the age groups. Amongst individuals between 15 and 24 years olds 70% have MP3 players, while in the age group above 65 years this proportion is only 1%.

Internet usage and age

As the digital divide debates showed, having access to the internet does not guarantee its use. So how many internet users are using this access in order to obtain music? More than one third of internet users in the sample claim they have downloaded music at least once (36%) – this translates into 15% of the above 15 years population as a whole, clearly a small minority.

We can now start to check our hypothesis about youth. In Figure 5 it is clear that the younger group downloads much more – about 40% of the individuals aged between 15 and 24 years old have already downloaded music. As we ascend through age groups the share of those downloading music declines –

26% of the 25 to 34 years age group, 8% within the 45 to 54 interval, less than 1% in the group aged 55 to 64 years and, we did not find anyone downloading among those more than 65 years old.

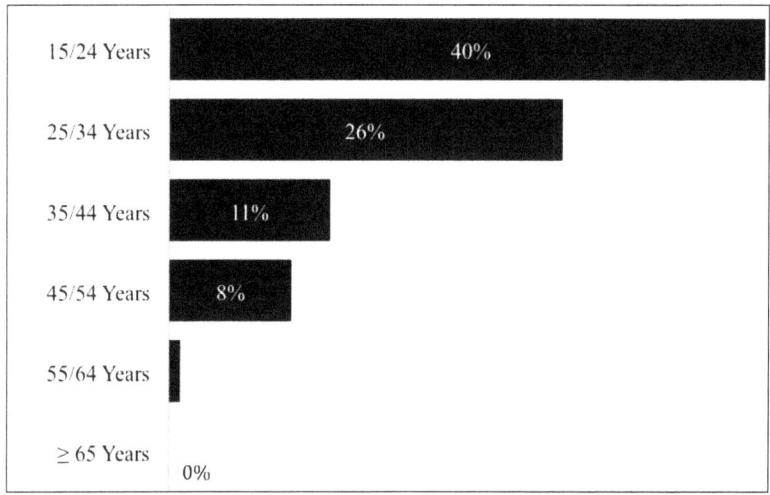

Figure 5: Music downloading by age groups.
Source: Network Society in Portugal, 2008. OberCom Base: 1039

However, the share of those downloading music may well be higher than the figures suggest and here we need to think about some of the processes shaping these statistics and what they might not show. First, the survey did not count people younger than 15 years (because of methodological difficulties involved when interviewing children). Had they been included this would have, most undoubtedly, swelled the overall number of consumers of music 2.0 and increased the predominance of 'youth' in this field. The second point is that one should reflect on the problematic nature of the interaction between interviewer and interviewee, bearing in mind that downloading commercial music without paying is an unauthorised practice with illicit contours, taking place in an antipiracy climate that stigmatises users, as seen by the growing number of law cases with legal convictions. All this combined may lead to some reluctance on the part of those interviewed to admit 'in public' to illegal downloading in a survey[71]. Hence the number of people downloading music is probably larger than the final figures suggest, and may well be larger, once again, for youth.

A second observation relates to diversity within youth. Here we need to abandon reified images that portray youth as an homogeneous mass of individuals. In practice, they have heterogeneous social and cultural back-grounds and biographies. Not all youngsters are the same, nor are all 'digital natives' the same. Thus we see that about half of this age group download music. That is higher than for other age groups but since half do not download, this reminds us of the uneven experience of music 2.0 even within this age group.

Use: downloading frequency

The various measures of use in the digital divide literature examine not only whether one uses the internet at all, but questions of the frequency of use, and in the case of music 2.0 one of the more obvious measures is frequency of downloading.

In figure 6, almost a third (29%) of those downloading music said that they download new music files less than three times a month and 15% did so between once every one and three months. But over half were more active than that: 25% said they downloaded on a monthly basis and 27% on a weekly one. At the more extreme end of the scale, the most active users, those downloading music daily, represented only 4%. This underlines how downloading for many people is a fairly regular practice. It also shows wide variations in experience, in terms of the frequency of downloading.

To put this into another perspective, if we compare this trend with the ones for the frequency of CDs purchase, digital musical consumption is much more prominent in everyday life. Only 5% of the people who purchase CDs said that they buy CDs on a weekly basis.

---

71 This hypothetical gap between reality and speech (social practices and social discourses) could be explored with other methodologies and research techniques, from *in loco* observation, focus groups, to computer monitoring and tracking devices.

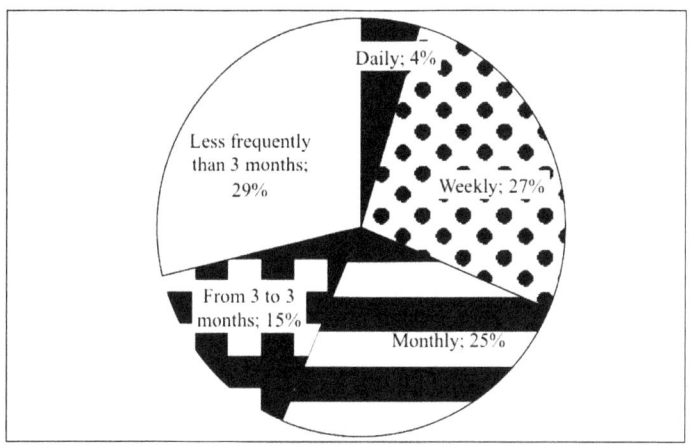

Figure 6: Frequency of downloading among those who download music.
Source: Network Society in Portugal, 2008. OberCom Base: 153

Downloading music and age

Let us return to our hypothesis about age, and internet use. From Figure 7 it is clear that the two youngest age groups who are also main internet users download much more than older internet users. Half of the youngest internet users, from 15 to 24 years old, download music, and 44% of the 25/34 age group downloads music also.

In order to generate a music 2.0 consumer profile, an analysis of multiple correspondence was conducted. Here, multivariate analysis was used to show visually a set of variables related to music 2.0 (music sharing, music downloading, internet use) alongside a block of socio-demographic variables: age, gender and educational level. Observing the bi-dimensional plan we can see that that some categories are spread apart, indicating a clear divide (see Figure 8).

165

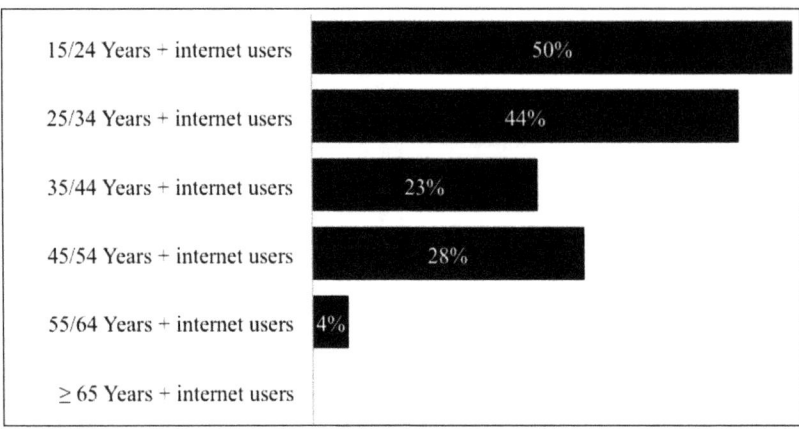

Figure 7: Music downloading by age group within internet users.
Source: Network Society in Portugal, 2008. OberCom Base: 423

On the other hand, the categories of sharing and downloading music via the internet are closer to each other, more associated with, younger people aged between 15 and 24 years, and those who have completed at least their secondary education – that is, it may also include those still attending higher education (but who have not yet completed it). Internet use is associated most with young-adults (25 to 34 years) and those with full educational qualifications of mid-to higher levels, but where the practice of downloading and sharing music is, even if present, less widespread. We can also observe that music downloading and music sharing practices are not in the identical space, reflecting the fact that users tend to download more than they share/upload.

Another group identified in this diagram are the non-internet users. This group is remote from downloading or sharing music practices, and its members are aged between 45 to 54 years. Moreover, we can observe the space of respondents with more advanced age (greater than 55 years) and with reduced educational resources (illiterate or with primary education level), are more distant from use of the internet and music 2.0 consumption. The diagram also shows a slight predominance of males using the internet and being consumers of music in its digital format.

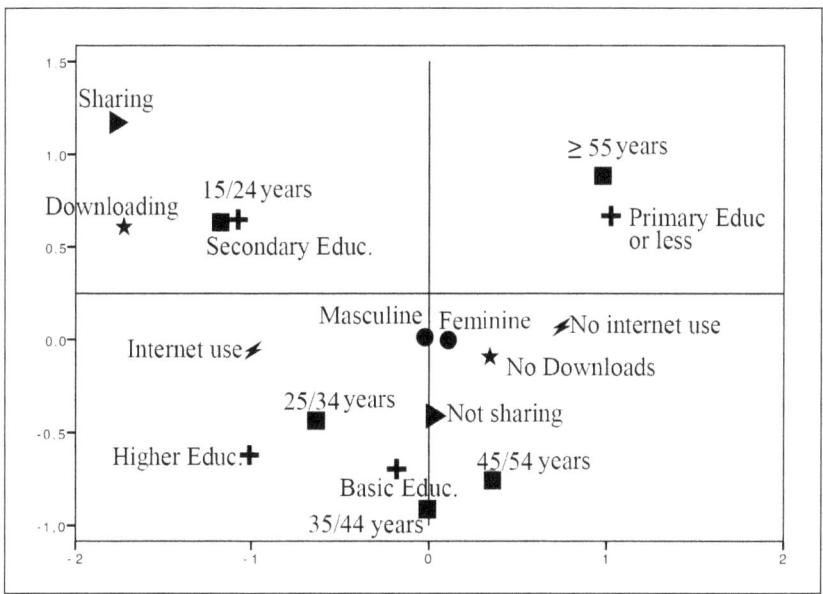

Figure 8: Music 2.0 profile.
Source: Network Society in Portugal, 2008. OberCom Base: 1039

Music 2.0 skills amongst internet users

Having followed the digital divide literature in terms of looking at access and use we can now consider another aspect identified in that body of work: online skills. In Figure 9, more than half (54%) of Portuguese internet users claimed that they did not know how to use P2P (i.e. peer-to-peer) software in order to share music or video content. We can also see that half of them (51%) did not know how to upload music online. Of all the skills that internet users were asked about, the most widespread was the ability to download music and videos (58%).

These data show first of all that the digital literacy required in order to obtain the benefit of digital music consumption is present only among, more or less, half of internet users and, secondly, this involves a slight predominance of downloading knowledge to the detriment of sharing music skills.

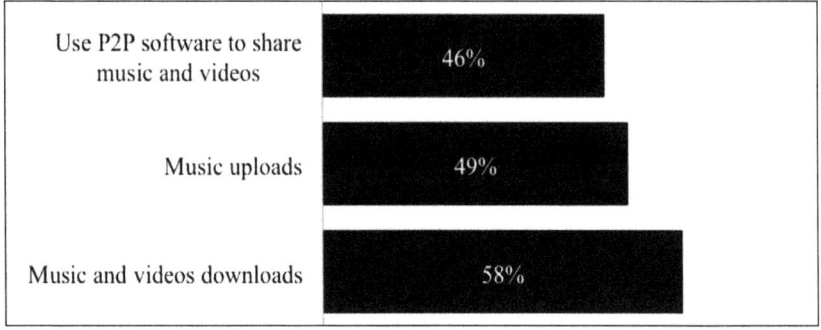

Figure 9: Sharing skills within internet users. Know-how.
Source: Network Society in Portugal, 2008. ObserCom Base: 423

Forms of participation in music 2.0 culture

Finally, besides the interest in non-users, the digital divide literature has tried to widen the discussion of 'use' to consider different aspects of the experience of the internet, including the extent of participation in the online world (Haddon, 2004). We can raise equivalent questions in the case of music 2.0. All the measures discussed so far have dealt with more traditional notions of consumption, in effect asking if people are consumers or not, and how much. But the spread of web 2.0 services more generally has also been accompanied by discussions on the rise of user-generated content. Since the 1980s many writers have suggested that we have to reconceptualise the existing idea of consumption as a passive process. In fact, in the 1970s the notion of 'prosumer' (Toffler, 1981) first gained a place in media theories and has been being refined by terms like 'produser' and 'co-creator' as one observes new forms of active and creative participation in the media field (Jenkins, 2006). These notions are used to represent those who produce what they or their peers consume and through that process are producing an alternative form of parallel wealth in a capitalist economy (Postigo, 2008).

The first step is to go beyond binary characterisations of this practice – producer versus consumer or professional versus consumer – and grasp the assorted roles open to users in this participatory culture more generally. This involves different levels of engagement where

> "user agency comprises different levels of participation, varying from 'creators' to 'spectators' and 'inactives'." (Dijck, 2009, pp.45-46).

In fact, Hargittai & Walejko (2008) show empirically that even if the barriers to producing and sharing content online are reducing, in practice only a small fraction of all users really capitalises on these new forms of participation to the full. Nor is such participation equally distributed among users. Some groups are under-represented, as we will see; women and older people, for instance. So there are gender gap and generational gap issues.

When applied to music, we can see some music fans in the forefront of the emerging phenomenon of MP3 blogs[72] that consist in personal web pages devoted to music sharing, and that requires a significant level of investment of time and effort. On the other hand we can see those that are only downloading music and taking advantage of these blogs without themselves ever posting an album online. In fact, those participating only as 'lurkers'[73] or 'leechers' represent the vast majority of users. A 2007 report from OECD (2007) claims that 80 percent of all users are passive recipients of content. So one can see that the number of very active, producing or sharing, users is disproportionally small.

But the picture is even more complex. If we now consider peer-to-peer networks we can imagine those users whose are actively sharing music – i.e. in this sense they are media providers – but that is all they do, and in part this is only is because of the ease with which the software application does it for them. But, besides this kind of activity we can think of those individuals who also comment online, creating social discourses around music, writing reviews that can stimulate buzz or hype and participating in other forms of viral marketing. For instance, when you share a YouTube link to your favourite song you are not actually uploading content but the process of mediation is nevertheless present.

Furthermore even by simply acting you can be adding value. Consumer behaviour can be stored and then used as a way of cataloguing/tagging content and capturing consuming practices using data-mining tracking devices. Therefore one should distinguish between different user roles, both as content providers and data providers (Dijck, 2009). And to be a data provider one only needs to act in a limited way – for example, by tagging a genre's song in last.fm or by buying on amazon.com.

So when we look at the actual evidence, can we see the participation gap (Jenkins, 2006) or participation divide (Hargittai & Walejko, 2008) in the Portuguese music scene? It seems so, since we can observe that almost a third

---

72 New forms of musical mediation are assisted not only by Peer-2-Peer techniques but also through the construction of personal pages that link to music files on external online hosting services (for instance: rapidshare.com, yousendit.com, megaupload.com, z-share.com and many others). Many of these pages specialise in particular music genres and attract a considerable amount of global users - like, for instance http://loronix.Blogspot.com, one entirely devoted to out-of-print Brazilian vintage vinyl.

73 The term 'lurking' refers to the kind of silent behaviour on message boards, forums newsgroup, file sharing or other interactive social system. (Nonnecke & Preece, 2003).

(30%) of internet users claimed to be sharing music content online, whereas only 14% said that they were actually uploading music online.

Music 2.0, musical diversity and the local. Uneven experiences.

Before summing up the contours of the music 2.0 experience, there remains one last question: what difference does the uneven experience make? What are the social implications? This has been an issue within the digital divide literature (Haddon, 2004) and it seems that this question is definitely both under-theorised and lacking empirical data in the case of music 2.0. Hence we can at this stage only find one potential measure and suggest some other possibilities.

One could argue that with music being available in this dematerialised format, its availability is not so tied to geographical contingencies as it was when it was distributed physically. So does this means that the possibility of accessing a global range of music is being realised in practice? More than half of Portuguese internet users (59%) agreed with the proposition 'The internet allows me to hear more bands and musical genres'. And, 90% of those downloading music thought that by hearing music online they had available to them a wider range of music options. This suggests that the musical portfolios among those downloading music are widening.

Consumers can now be more aware of music news, can listen to audio snippets, browse tops and reviews, and download within a few seconds a tune just finished and released almost in real time. Or they can be at home and request an album to be delivered by normal mail. We must not forget that before the digitisation of music and network distribution one was more constrained to local catalogues. So this ease and variety of access to content, independent from spatial constraints, may be felt as a form of empowerment for some music consumers, especially those living in peripheral areas. In the days when music was distributed in physical formats (CD's or vinyl), buying foreign music usually implied paying much more – since it was imported. With online sharing, rates are more standardised, so it is cheaper and more convenient to download music than to buy a CD in a local store. And one must bear in mind that the range of physical records available locally (where one lives) is smaller than the catalogues available online. How does this affect the user experience? Would those downloading music feel deprived if they could not access music online? Do these changes in music distribution and access suggest an eroding Anglo-Saxon musical hegemony? Is there now more space for local musical traditions to flourish since they can be more easily mediated?

On the other hand, even with music 2.0 there are still spatial constraints. For instance, someone with an Internet Protocol (IP) address[74] located in Portugal cannot access some music services located in the United States of America, whether via online radio or in the form of a digital music store, because of different legal frameworks regarding authorship, copyright and public performance issues. As the IP configuration sets a geographical reference, some online services are not available in some countries, the catalogue may differ, and there are cross-country price variances. So geography still makes a difference and the local is still socially relevant in terms of social inequalities in different user experiences.

Also, there are still infrastructural differences and social impacts concerning internet speeds (Anderson, 2008 and chapter five in this book). According to the online service Speed Test[75] it will, for example, be faster on average to download an album in Sweden than in Portugal (13,14 Mb/s VS 8,39Mb/s), and it will be much faster to upload it in Sweden where average upload speed rounds 4.66 Mb/s.

Lastly, the language barrier comes to mind. English is becoming more and more the global lingua franca, the vehicular language structuring online relations and communications. So a user who does not understand English will be confined to a small range of online shops and opportunities.

Conclusions

Digitisation allows content complementarity between different media and technologies, so some have spoken of music 2.0 amounting to a 'revolution' in digital entertainment as a cross-media system. The historical account of the evolution of music reminds us of how music 2.0 is also part of a continuum of developments that has led to changes in music consumption over time, even if some regard these web 2.0 developments as constituting a major shift.

Having presented, both in theoretical discussions and in empirical data, some sense of the importance of music in everyday life (at least in Portugal), the chapter then explored the contours of music 2.0 consumption. It first built on the digital divide literature on internet access, looking not only at those who have no access, but also those with different types of access. The caveat, discussed in relation to MP3 players, is that those with no access can still benefit from music 2.0 by proxy access through others. Overall, the figures may be for Portugal, one of the European countries with relatively fewer people online, but the principles of what to measure could be applied more generally.

---

74 A number assigned to a device that serves for host identification and location addressing.
75 Available at http://www.speedtest.net [Accessed 16 September, 2009]

Drawing on the digital divide literature with regard to the issue of use, this chapter examined both those who have downloaded music as well as the frequency of downloading. These sections also specifically examined the profile of users, testing a hypothesis about the importance of young users, but also showing how much variation exists between those young users, in order to avoid stereotyping the music experiences of youth. Data looking at the level of skills people have that relate to music 2.0 also revealed considerable variation, especially relevant in relation to the next questions about forms of participation online.

The digital divide literature also looked at an expanded version of the experience of the internet, beyond mere use and including participation online. In the case of music 2.0 this chapter has built upon the specific discussions of what it means to participate in web 2.0 more generally – demonstrating the different forms of music 2.0 participation and what a participation gap might look like.

Finally, although this has inevitably been more speculative, we have least started to consider what might count as measures of the consequences of access to and use of music 2.0, and of different levels of access to it.

Ideally, in keeping with the spirit of this book, it would have been useful to explore how the contours of music 2.0 charted here vary across countries. However, the problem is that these kinds of issues are only now beginning to enter into political and research agendas. If web 2.0 media research generally is still relatively scarce, music 2.0 research is in an embryonic state and this is particularly so for internationally comparable data in this field.

Bibliography

Anderson, B., 2008. The social impact of broadband household internet access. *Information, Communication & Society,* 11 (1), pp.5-24.

Appadurai, A., 1996. *Modernity at large: cultural dimensions of globalization.* Minneapolis: University of Minnesota Press.

Asante, M., 2008. *It's bigger than hip hop: the rise of the post-hip-hop generation.* New York: St. Martin's Press.

Axelsen, D., 1997. *Listening to recorded music: habits and motivation among high-school students,* Gothenburg: Acta Universitatis Gothoburgensis.

Barfe, L., 2005. *Where have all the good times gone? The rise and fall of the record industry.* London: Atlantic.

Beer, D., 2008. The iconic interface and the veneer of simplicity: MP3 players and the reconfiguration of music collecting and reproduction practices in the digital age. *Information, Communication & Society,* 11 (1) pp.71-88.

Benjamim, W., 2008. *The work of art in the age of mechanical reproduction.* London: Penguin

Bull, M., 2007. *Sound moves: iPod culture and urban experience.* London: Routledge.

Chang, J., 2005. *Can't stop, won't stop: a history of the hip-hop generation.* New York: St. Martin's Press.

Connell, J. & Gibson, C., 2002. *Sound tracks: popular music, identity, and place*, London: Routledge.

DeNora, T., 2002. The role of music in intimate culture: a case study. *Feminism Psychology*, 12, pp.176-181.

Dijck, J., 2009. Users like you: theorizing agency in user-generated content. *Media, Culture and Society*, 31 (1), pp.41-58.

Dimitriadis, G., 2001. *Performing identity/performing culture: hip hop as text, pedagogy, and lived practice.* New York: Peter Lang.

Du Gay, P. et al., 1997. *Doing cultural studies: the story of the Sony Walkman.* London: Sage.

Frith, S. 1996. Music and identity. In: S. Hall & P. Du Gay, eds. 1996. *Questions of cultural identity*, London: Sage.

Gauntlett, D. & Horsley, R. ed. 2004. *Web studies.* 2nd ed. London: Arnold.

Gelder, K., 2007. *Subcultures: cultural histories and social practice.* New York: Routledge.

Giddens, A., 1990, *The consequences of modernity.* Cambridge: Polity.

Gilbert, J. & Pearson, E., 1999. *Discographies: dance music, culture, and the politics of sound.* London, New York: Routledge.

Hargittai, E. & Walejko, G., 2008. The participation divide: content creation and sharing in the digital age. *Information, Communication & Society*, 11 (2), pp.239 – 256.

Haddon, L., 2004. *Information and communication technologies in everyday life: a concise introduction and research guide.* Oxford: Berg.

Harrison, A., 2000. *Music: the business.* London: Virgin.

Hodkinson, P. & Deicke, W., 2007, *Youth cultures: scenes, subcultures and tribes.* New York; London: Routledge.

Jenkins, H., 2006. *Confronting the challenges of participatory culture: media education for the 21st Century.* Occasional paper for the MacArthur Foundation. [Online] Available at: http://digitallearning.macfound.org/atf/cf/%7b7e45c7e0-a3e0-4b89-ac9c-e807e1b0ae4e%7d/jenkins_white_paper.pdf [Accessed 15 October 2009].

Katz, M., 2004. *Capturing sound: how technology has changed music.* Berkeley: University of California Press.

Kendall, L., 2008. Beyond media producers and consumers: online multimedia productions as interpersonal communication, *Information, Communication & Society*, 11(2) pp.207-220.

Leonhard. G., 2008. *Music 2.0*. Hämeenlinna: Gerd Leonhard [Online] Available at: http://www.music20book.com [Accessed 23 September 2009].

Levy, S., 2007. *The perfect thing: how the iPod shuffles commerce, culture, and coolness*. New York: Simon & Schuster Paperbacks.

Loader, B., 2007. *Young citizens in the digital age: political engagement, young people and new media*. Abingdon: Routledge.

Longhurst, B. 2007. *Popular music and society* 2nd ed. Cambridge: Polity.

Nonnecke, B. & Preece, J., 2003. Silent participants: getting to know lurkers better. In: D. Fisher & C. Lueg, eds. *From usenet to co webs: interacting with social information spaces*. Dordrecht: Springer. pp. 110–132

Organisation for Economic Co-operation and Development (OECD), 2007. *Participative web and user-created content: web 2.0, wikis and social networking*. [Online] Available at: http://213.253.134.43/oecd/pdfs/browseit/9307031E.pdf [Accessed 20 February, 2009].

O'Hara, K. & Brown, B., 2006. *Consuming music together: social and collaborative aspects of music consumption technologies*. Dordrecht: Springer.

Postigo, H., 2008 Capturing fair use for the YouTube generation: the Digital Rights Movement, the Electronic Frontier Foundation and the user-centered framing of fair use. *Information, Communication & Society*, 11(7) pp.1008-1027.

Prensky, M., 2005. Listen to the natives. *Educational leadership. Learning in the digitalge*. 63 (4) pp.8-13

Shuker, R., 2008. *Understanding popular music*. 3rd ed. London: Routledge.

Styvén, M., 2007. The intangibility of music in the internet age. *Popular Music and Society*, 30(1) pp. 53-74.

Théberge, P., 2004. The network studio: historical and technological paths to a new ideal in music making. *Social Studies of Science,* 34 (5), pp. 759-781.

Toffler, A., 2006 *The third wave*. New York, Bantam

VA, 2000. *Consumers call the tune: the impact of new technologies on the music industry*. A report by the New Technology Group of the Music Industry Forum. London: Department for Culture, Media and Sport.

Vickery, G. & Wunsch-Vincent, S., 2007. Participative web and user-created content: web 2.0, wikis and social networking. Paris: Organisation for Economic Co-operation and Development.

Webb, P., 2008. *Exploring the networked worlds of popular music. Milieu cultures*. New York: Routledge.

Wurtzler, S., 2007. *Electric sounds: technological change and the rise of corporate mass media*. New York: Columbia University Press.

Yoshikawa, M., 2007. Foreseeing the future lifestyle with digital music: a comparative study between mobile phone ring tones and hard-disk music players like iPod. In: S. Van Der Graaf & Y. Washida, eds. *Information communication technologies and emerging business strategies*. Hershey, PA: Idea Group Pub, pp.59-75

Leslie Haddon

Chapter nine. Methodological issues in the cross-national analysis of contextual data

Introduction

This chapter aims to reflect upon the methodological strategies and issues involved in conducting a cross-national analysis of a broad range of internet studies. In particular, it looks at the possible options for analysis when a study involves many countries (rather than just two or three). The type of data considered here is the background information that one might want to know about societies in order to contextualise any studies conducted in them, e.g. an overview of the national social structures, the countries' legal systems, the nature of media coverage of various issues, recent histories of pertinent events, how research is organised in these countries, etc. While, it is possible to count quantitative indices, much of what we examine under the heading of 'contextual factors', would often be considered more qualitative in nature. Hence there is a challenge in conducting comparative analysis using such data.

The chapter is based on two traditions. It is firstly written in the spirit of sharing insights into what goes on behind the scenes in an ICT study, where dilemmas, problems of classification and even strategies that are abandoned are not necessarily ever fully discussed in the accounts that are eventually published and thus in the public domain. Therefore, it is the same genre as writings reflecting upon the dilemmas in ICT design (Limonard and De Koning, 2005) or outlining the problems of classification systems relating to what counts as 'rural' and 'urban' for mapping the adoption of ICTs (Gilligan, 2005)[76]. This tradition of reflection on how analyses were managed and how methodological and analytical decisions have been made, and with what implications, dates back to the 1980s, although the discussion at that time was not specifically looking at information and communication technologies (ICTs) (e.g. Bell and Roberts, 1984; to some extent Roberts, 1981). The other relevant literature is the small but growing one on cross-cultural analysis (e.g. Haantrais & Steen, 1996), which has already established many of the issues, but which has for the most part not specifically looked at ICTs, with some notable exceptions (Blumer, et al, 1992; Livingstone & Bovill, 2001). For a previous appraisal of the literature on cross-national studies, see Livingstone, 2003.

---

76 On this issue, see more generally the volume Haddon, et al., 2005.

Background: The EU Kids Online study

The *EU Kids Online* project was a 21-country[77] study evaluating European research on children's experiences of the internet. The project, funded by the EC's *Safer Internet plus Programme,* collected and examined information about existing studies in the countries concerned (see the book from the project: Livingstone & Haddon, 2009). One sub-project identified the patterns of studies both across and within the countries participating in *EU Kids Online* and on this basis drew attention to the need for further research in certain areas (Staksrud, et al., 2007; Donoso, et al., 2009). Another sub-project conducted a methodological literature review and developed a Best Practice Guide for researching children, researching internet use and conducting cross-national comparisons (Lobe, et al., 2007; Lobe, et al., 2008; Lobe, et al., 2009). But it is two of the other strands of the project that will be re-examined here:
- a comparative evaluation of the actual data on children's experience of the internet (Hasebrink, et al., 2008).
- an analysis of factors shaping why certain types of research on children and the internet are conducted and why this varies across countries (Stald & Haddon, 2008[78]).

More generally, across the *EU Kids Online* project the various strands were simultaneously dealing with the wider methodological challenge of how, systematically, to manage cross-national comparisons[79]. Hence, some of the procedures and decisions behind two reports noted above are here re-examined not for their substantive findings, but as illustrations of some of the issues faced at the stage of data analysis. These two different reports were chosen, with different foci, because of what is of interest in the common challenges faced when trying to examine the type of contextual factors noted above. We will now outline in more what these included.

The focus on the first of these sub-projects, the Hasebrink, et al. (2008) report, was on where, and to what extent, there are European commonalities or differences regarding children's online experiences, risks and opportunities. What common European responses and patterns have been identified and what factors explain these? The contextual factors that the report considered were:

---

77 Austria, Belgium, Bulgaria, Cyprus, Czech Republic, Denmark, Estonia, France, Germany, Greece, Iceland, Ireland, Italy, Norway, Poland, Portugal, Slovenia, Spain, Sweden, the Netherlands and the United Kingdom. For more details see www.eukidsonline.net [Accessed 29 December 2009]. The researchers involved in the coordination of the study are Sonia Livingstone, Leslie Haddon, Panayiota Tsatsou and Ranjana Das.

78 A shorter published version appeared as Haddon & Stald, 2009a.

79 The justification for focusing on nations, given that there is some debate about this strategy, is provided in Hasebrink, et al., (forthcoming).

- The internet's diffusion
- Internet safety tools and initiatives
- Media content for children (broadcast and online)
- Internet regulation and promotion
- The role of the government and the regulator
- The influence of NGOs
- Public discourses about the internet including media coverage of children and the internet (which was a separate empirical sub-project within the broader project – Haddon & Stald, 2009b; Ponte, et al., 2009) and the role of NGOs in shaping these discourses.
- Wider national values and attitudes
- The education system (including internet access and use within schools)
- Wider country-specific factors (e.g. social change, inequalities, urbanisation, work and social class, free speech and censorship, migration and cultural homogeneity, the role of the state, the extent to which English as a language is understood and the extent to which a children's 'bedroom culture' exists)

The focus of the second sub-project, in the Stald and Haddon (2008) report, was on the social shaping of research: what social factors influence why certain research on children and ICTs takes place and, as the comparative element, why do different amounts of research exist in different countries and why are some research questions followed up in some countries more than others? The contextual factors considered were:

- The size of the national research base, the activities of different disciplines
- Institutional processes (e.g. the national histories of social science and related research in general)
- Funding sources (e.g. Government, commercial)
- Political initiatives (e.g. internet awareness campaigns, education initiatives)
- Public discourses (e.g. media coverage of children and the internet; whether there had been specific and important events in this field)
- Particular debates (e.g. about the commercialisation of childhood).

Steps in the analysis

In both sub-projects, the way in which the material was organised was the same (see Hasebrink, et al., 2009, forthcoming for a more detailed outline and reflection on procedure). The above areas of interest were organised into a set of questions to be answered for each of the countries participating in the *EU Kids Online* project. For both parts of the project, each of the national teams then wrote their national reports answering the questions related to the issues outlined above. Subsequently, there was a division of labour such that different researchers looked at the different contextual factors across countries. For example, one person or a group of people would specialise in looking at all of the material on, say, the role of government and the regulator in the first sub-project, or funding sources in the second sub-project.

It is useful to reflect a little further on the processes at work here. Setting up a template for the national reports is something akin to writing an international questionnaire (with many open-ended answers). Arguably one can ask for reasonably sophisticated answers, and the teams do have time to consult with colleagues, mobilise supporting evidence, etc. Nevertheless, one has to find a form of words in the template that enables somewhat comparable answers. The questions were discussed by the groups involved in the two respective sub-projects and in this sense were piloted. While many of the questions tried to be as specific as possible in order to 'manage the theoretical diversity' represented with the project (Swanson, 1992, p.25), the national teams had to decide how to address the questions as regards their own countries, which introduces some variation when deciding how to answer. Those project members analysing the various national commentaries and evidence relating to the different contextual factors then had to work out what strategies they would use to manage the feedback they received within these national reports.

The final point to make by way of scene setting relates to the two main approaches at work in conducting this data analysis and presenting the material. Both work packages, in their different ways, used Kohn's (1998) framework for cross-national analysis. Specifically, they used two out of his four modes on cross-national comparison: his notion of nations as units of analysis versus nations as contexts for study. In the case of countries as units of analysis, the aim was to try to explain similarities and differences between countries – i.e. this was the comparative element noted above. But at times, both sub-projects looked at nations as case studies, pooling the data from the different countries in order to have sufficiently rich material to describe a particular phenomenon of interest. For example, the Hasebrink, et al. (2008) report brought together studies from the different countries (not specifically the contextual factors outlined above) in order to evaluate more general hypotheses about the relationship between children and the internet (Hasebrink, et al., 2008). In the Stald and Haddon (2008) report for some of the topics the aim became to show how a particular process worked, say, within research institutions, where examples pooled from the different countries could illustrate how this could operate in a variety of slightly different guises. In the Hasebrink, et al. (2008) report this was a more explicit, intentional strategy from the start for some issues. In the Stald and Haddon (2008) report, this strategy emerged given the nature of the information that the national teams could supply – i.e. while it was insightful, it was sometimes not possible to use this material to compare countries.

Strategies in the analysis

The continuum from quantitative to qualitative

The first observation about the analytical process comes from Stald and Haddon (2008) report, more specifically relating to the institutional influences shaping what research takes place. Different questions produced answers that, especially at the point of data analysis could be used more of less quantitatively or qualitatively, almost on a continuum,

Amongst a range of questions asking about traditions and histories of research (e.g. about the degree to which qualitative traditions were established in different countries, the dates when mass media were first researched), one question asked when the first internet studies appeared. The list of dates, or at least periods, was tabulated and it was found that the national differences in the timing of research roughly correlated with country levels of usage by children, one of measures being used for other parts of *EU Kids Online* analysis (but which we know, in turn, reflects general internet penetration rates). Hence here was an example where one could tentatively suggest that it looks as if the timing of research on children and the internet more or less reflected the actual take up of the internet. Here was a case of comparing nations as units.

There were then various questions about university procedures when applying for research, including whether national regulations exist about what cannot be researched as regards children, whether there are some fixed stages that all research proposals have to go through, and whether proposals have to be checked by the applicants' institution/department before they can proceed. These questions aimed to explore how complicated it is to organise research and whether national variation might exist. These could be tabulated because the answers were often 'yes' or 'no', even if national teams frequently added a few further qualifications and exceptions. In practice, these questions produced a picture of what was common practice (e.g. few 'hard' rules, but some 'soft' ones), showing a little country variation but nothing that could be systematically related to particular patterns of national research.

The same type of analysis emerged from questions asking whether Government Ministries ask for certain types of research to be conducted (e.g. via research councils) and whether there were pressures to collaborate with industry (both of which might in principle have directed research in certain directions rather than others). What might have been anticipated but was certainly discovered in practice is that in comparison to questions about procedures these generated far more wordy explanations. Tables of answers were supplied in the Stald and Haddon (2008) report, points were made about common patterns and trends (when analysing nations as units). However, on balance it made more sense to use this material in the form of nations as case studies, combining the descriptions to develop qualitatively a more complete understanding of different levels on which, or manner in which these same type of pressure were experienced (Stald & Haddon, 2008).

The last example was a question about whether there was general pressure on university employees to conduct research? This clearly proved to be an invitation for national teams to explain the myriad ways in which such pressures operated from those related to the way in departmental budgets worked, through factors affecting an individual's career progression to departmental expectations about the number of academic publications one was supposed to produce. Hence the decision was made not to try to count these 'pressures', but instead use the material, including quotations form the national reports, to explore this set of contemporary social incentives to conduct research than in the past.

The point of these four examples is that they illustrate how this contextual material could be used in a combination of quantitative and qualitative ways, and the logic of using the nation as unit or as context of study depended partly on the form of the question (e.g. asking for a date, asking for a yes/no answer, inviting longer answers) but partly also on what national teams actually wrote.

Counting issues

One initial area of interest was whether the amount of research on children and the internet reflected the overall amount of research that takes place within a country: i.e. do countries that in general have a good deal of research also have a good deal specifically on children and the internet? But at the planning stage it was clear that it would be difficult to measure this overall level of research. As a proxy, the only measurable unit where data might be available in all the countries related to the size of what was termed its 'academic base' (i.e. the number of academic institutions)[80]. Since the interest was in research institutions (rather than purely teaching ones), this was operationalised in terms of counting the number of universities, since the latter are often listed somewhere for each country. That said, the figures have to be taken with caution and the aim was to give an idea only of relative size of the academic base – the only example where we had some figures to demonstrate this point is Estonia, which had 11 universities but 75 registered academic research bodies.

In practice, even mapping the universities was not straightforward. Apart from bodies called 'universities' in France there are various Grandes Écoles and

---

80 The main single alternative source to academic research was commercial research, accounting for only 18% of all studies, varying by country, and problematically this research is not always publicly accessible.

Grandes Établissements, which are universities except in name –so these were included. In contrast, the final British figures excluded the 'university colleges' (more teaching oriented), while the numbers had to be expanded to account for bodies like 'London University' because this is an umbrella organisation that effectively includes a number of universities in their own right (like the London School of Economics). In other words, if there was a reasonable rationale the base figures could be adjusted. There proved to be a high correlation, by and large and with some exceptions, between the number of universities in a given country and the size of the population base, which could in this case be demonstrated graphically. Of more interest for the project, although the correlation was less strong, and with more exceptions, the larger the academic base the more studies there were of children and the internet.

The problems of counting worsened in the case of disciplines. The first *EU Kids Online* report, on research gaps, had already noted that sometimes it was difficult to decide the discipline informing a specific piece of research, especially when research was interdisciplinary or, in many cases, market research (Staksrud, et al,. 2007). Nevertheless, it was decided to experiment with some potential lines of analysis, if only to see whether they looked productive. From that first report it was clear that there is a fair amount of research conducted within Education and Psychology departments, but these disciplines as well as Sociology are established in most universities in most European countries and so counting these would not differentiate counties for comparative purposes – the result would more or less replicate the figure for the academic base. One possible hypothesis was that Media Studies and Communication Studies might be disciplines more likely to conduct research in this field, but as newer disciplines they might not be so established in all countries – and this would be even more true of newer subjects like 'New Media', 'IT and Society' and 'Informatics'. In other words, one can ask at least whether the prevalence of these departments could help to explain some country variation in the amount and type of research.

When trying to ascertain the number of Media or Communications Studies departments there were in each country there were a number of practical issues. The first one, relatively straightforward, related to names. In France, the subject matter of Media and Communications Studies is usually researched under the heading 'Science de l'Information et de la Communication' while in Denmark what is in effect Communication Studies is sometimes called 'Information Studies'. In these known cases, it was possible to allow for this when making calculations. More problematic was the fact that many Media Studies and sometimes Communications Studies departments were very practically oriented (e.g. in Germany, and very often in the Czech Republic), teaching production skills or journalism. Up to a point this could also be allowed for, not counting departments whose name indicated that they were clearly oriented to, say, journalism, or where the *EU Kids Online* national teams knew how particular departments worked.

However, more detailed comments made in national reports showed the weaknesses in even the adjusted data. First, media and communications may be studied and researched in departments not using that name. For example, in Spain Media and Communications Studies do not exist, while empirical research on audience behaviour, for example, is likely to appear within Sociology and Social Psychology[81]. Meanwhile within Belgium, in Flanders, Media Studies is a discipline in own right whereas in French speaking Wallonia the subject matter is often taught within Social and Political Sciences. Second, when separate Media and Communications Studies departments exist their orientation can then depend on the larger faculty within which they are located. For example, in Denmark, if they are located in the Humanities they have a more philosophical, literary and aesthetic orientation but when located within the Social Science faculties they are more empirically oriented (which is of more interest for our examination of internet research). In Germany Communication Studies is more often located in the social sciences whereas Media Studies is more often linked with film analysis and positioned in the humanities. In Portugal Media Studies is more oriented to textual and visual analysis rather than 'reception studies' (the empirical studies of interest in this report). In Italy Media Studies can be taught within the Humanities, Arts, Social Sciences or Education.

From the figures it was only possible at best to demonstrate that some of those countries where Media and Communication Studies are well established in universities appear to produce more studies on children and the internet – such as Belgium, Sweden and the UK. But while it had been important to ask the question about the influence of disciplines, in this case the chief discovery probably related to the difficulties of counting[82]. In fact, that problem proved to be even worse when trying to count Mew Media, IT and Society and Informatics departments and so that particular strand was abandoned altogether in the light of the feedback in the national reports.

Grouping countries

Given the large number of participating involved, one strategy adopted was to organise the countries involved into groups when considering how to evaluate

---

[81] Moreover, the internet tends to be studied by academics based in philosophy, discussing more general effects on society than conducting empirical studies of internet behaviour.

[82] There is a related discussion of comparing European official statistics, in this case of employment categories, when the terms used mean different things in different countries - see Desrosières, 1996.

them in relation to any particular question (e.g. internet diffusion, media coverage and educational levels in the Hasebrink, et al. (2008) report; funding sources and how many Media and Communications university departments they had in the Stald and Haddon (2008) report. In fact, this was addressed more systematically in Hasebrink et al report, because those responsible for the analysis of different contextual factors were specifically asked if it was possible to create meaningful clusters of countries pertinent to the interests of the *EU Kids Online* project. That said, occasionally there were discussions of exceptional individual countries, sometimes outliers on some scale, if this was useful for raising issues. For example, Denmark had strikingly different media coverage from most other countries taking part in a 14-country press analysis (see below) and so it was a useful case study for discussing the processes that might be at work. UK and German research had substantially more commercial funding compared to other countries, which was noted since it played a part in boosting the number of studies in those countries.

Although this chapter has stressed the point that much of the contextual information was qualitative, it did at times draw on existing pan-European (or even global) statistical sources as a basis for comparison, especially in the Hasebrink, et al. (2008) report. For example, the section discussing internet diffusion drew on Eurostat figures and the one charting the success of Governments in promoting ICTs could use one of the measures from the Network Readiness Index. A section looking at whether laws were well developed and enforced could cite the results of the Executive Opinion Survey used by the World Economic Forum. Researchers examining the social values prevalent in different countries could utilise the European Values Survey, (in this case re-analysing the data using factor analysis) and educational attainment figures came from the OECD. Material from some existing reports was also used where they had already complied information e.g. countries had been grouped by the age limit at which pornography is considered to be 'child pornography'.

Since there was an absence of suitable material showing media coverage in the field of children and the internet, *EU Kids Online* conducted its own media content analysis of press stories in this field. This provided the basis for the further clustering of countries, e.g. according to the degree to which they covered the different types of risks related to online content (e.g. aggressive content, sexual content), contact with strangers online and the conduct of children themselves on the internet (e.g. cyberbullying).

Lastly, the qualitative material could itself sometimes by used as the basis for scales by which countries could be classified. For example, in one section countries were grouped according to whether national Internet Service Providers (ISPs) played an active role in safeguarding safety online, whether they simply offered safety packages or whether they provided (almost) no warnings or advice. And in another section, countries were grouped according to whether their NGOs had been active and influential, active but not influential or not very active at all. Although allocation to such groupings required a substantial amount of subjective judgement, it was based on evidence cited in the national reports.

Meanwhile, since the *EU Kids Online* project had developed a database of entries describing the various European studies, in the Stald and Haddon (2008) report it was possible to chart their funding arrangements and organise classifications of countries accordingly (e.g. building typologies according to different combinations of funder, such as countries where public funding of research predominated).

At one level, these clusters were useful descriptively for drawing attention to where patterns existed, especially ones that might be pertinent for the area being studied. To develop some of the examples listed above, it became clear from the clustering process that there are different degrees to which laws are developed and enforced in the different countries, that media coverage of risks varies by country and that national NGOs can be more or less active and influential, all of which, in principle, could be relevant for understanding country variation in risk perceptions and behaviour.

However, the next stage involved using the clusters more analytically, asking to what extent the classification of countries on a particular dimension related to some other pattern that had been examined, such as the take up of the internet by children, the degree of risk in certain countries, or whatever. Occasionally there were links, if we now take some examples from the Hasebrink et al (2008) report. There was a high correspondence between cultural values dominant in countries and the overall country classification based on children's internet use and the degree of online risk. Another clustering process showed that the higher the general education of a country, the higher its children's internet use. In general, in countries where the internet is less common, more efforts are made to promote internet use, while once the internet becomes more common, risk awareness and then literacy initiatives become more prevalent. Finally, the more internet users countries have, the more legislation they have regulating activities on the internet.

We can take comparable examples from the clustering process in the case of the funding analysis in the Stald and Haddon (2008) report. The balance of funding sources varied across European countries, but in general various 'public financiers', especially national governments and the EC, were the most important sources of research money. That said, very different national funding arrangements were capable of generating large amounts of research – in this sense, there was no one type of funding structure, or balance of funding from different sources, which produced the most research. And as another 'negative' finding, or lack of correlation, the sources of funding did not appear to affect the topics being researched either For example, countries with a large amount of public financing did not necessarily produce much research on risks issues, and those with substantial commercial funding, did not produce research limited to such things as access and usage. It seems that different types of funder can actually have quite diverse interests, varying by country.

This process of clustering countries also provided ideas for further hypotheses, again more systematically developed in the Hasebrink et al (2008) report because those conducting the analysis of the various contextual factors were specifically asked to propose possible hypotheses (which might be tested in future research). For example, having noted different types of risk get more and less coverage in different countries' press, one hypothesis would be that in countries where there was more press coverage of content risks online (i.e. what problematic content children might encounter online), there would be more parental concerns about this issue (and the equivalent for contact and conduct risks). The point is that once the media variation is recognised, one hypothesis will be that, depending on country, national media will sensitise the public to different issues. Of course, even if this process were to be occurring, this may simply be reflected in data about parental attitudes since the media provide just one public discourse; awareness raising campaigns may work in a different direction, for instance. To take another example, one hypothesis that emerged from this process was that the presence of information and guidelines about online safety in ISPs' websites may well have a positive effect on children's behaviour and attitudes regarding online safety issues.

Conclusion

It is appropriate to remember the particular conditions under which this analysis took place, as specified at the start. Although at times particular narrow hypotheses were explored, this was in general a very ambitious project often exploring what factors might have a bearing upon the objects of study in the two sub-projects: children's internet experience and the shaping of research. Hence the contextual questions asked were often very demanding, requiring national teams to search for evidence. It is perhaps not surprising that in many cases there were at least two and sometimes several team members per country given the nature of this workload. In that respect, the procedures by which the material was assembled was not necessarily akin to that followed in some other cross-national studies. But in addition we have to take into account the sheer number of countries involved. In fact, there was a 'pilot report' involving just three countries – Poland, Portugal and the UK – whose aim was to establish some of the principles of analysis to be rolled out in the full 21-country study (Hasebrink, et al., 2007). While it achieved this goal, it was also clear that comparing three countries is a very different exercise, a very different form of analysis, from comparing many countries – for example, not involving the clustering process outlined above which suits the larger study. With these qualifications in mind about the generalisability of the points raised to other cross-national research, this chapter has shared some of the challenges the were faced in the analysis of contextual data within the *EU Kids Online* project, to indicate the basis for decision-making during data analysis and to illustrate the types of analysis generated in this process.

Bibliography

Bell, C. & Roberts, H. eds. 1984 *Social researching: politics, problems, practice*. Routledge and Kegan Paul: London.

Blumer, J, McLeod, J. & Rosengren, K., eds. 1992. *Comparatively speaking: communication and culture across space and time*. Newbury Park: Sage.

Desroières, A., 1996. Statistical traditions: an obstacle to international comparisons? In: L. Hantrais, L. M. Steen, M., eds. *Cross-national research methods*. London: Pinter, pp.17-27.

Donoso, V. Ólafsson, K. & Broddason, T., 2009. What we know, what we do not know. In: S. Livingstone & L. Haddon, eds. *Kids online. Opportunities and risks for children*. Bristol: Policy Press, pp. 19-31.

Gilligan, R., 2005. Questioning the "rural" adoption and use of ICTs. In L. Haddon, et al. eds. *Everyday innovators: researching the role of users in shaping ICTs*. Dordrect: Springer, pp.155-167.

Haddon, L. & Stald, G., 2009a. Cultures of research and policy in Europe. In: S. Livingstone & L. Haddon, eds. *Kids online. Opportunities and risks for children*. Bristol: Policy Press, pp.55-70.

Haddon, L. & Stald, G., 2009b. A comparative analysis of European press coverage of children and the internet. *Journal of Children and Media*, 3 (4), pp. 374-93.

Haddon, L, et al., eds. 2005. *Everyday innovators, researching the role of users in shaping ICTs*. Dordrect: Springer.

Haantrais, L. and Steen, M., eds. 1996. *Cross-national research methods*. London: Pinter.

Hasebrink, U. Livingstone, S. Haddon, L. Kirwil, L. & Ponte, C., 2007. *Comparing children's online activities and risks across Europe: a preliminary report comparing findings for Poland, Portugal and UK*. [Online] Available at: http://www.lse.ac.uk/collections/EUKidsOnline/ Reports/Default.htm [Accessed 29 December 2009].

Hasebrink, U. Livingstone, S. & Haddon, L., eds. 2008. *Comparing children's online opportunities and risks across Europe: cross-national comparisons for EU Kids Online* [Online] Available at: http://www.lse.ac.uk/collections/EUKidsOnline/Reports/Default.htm [Accessed 29 December 2009].

Hasebrink,U. Ólafsson, K. & Štětka, V., 2009. Opportunities and pitfalls of cross-national research. In: S. Livingstone & L. Haddon, eds. *Kids online. Opportunities and risks for children.* Bristol: Policy Press, pp.41-55.

Hasebrink, U. Olafsson, K. & Steka, V. (forthcoming) Commonalities and differences. How to learn from international comparisons of children's online behaviour. *International Journal of Media and Cultural Politics.*

Kohn, M. L., 1998. Introduction. In: M. Kohn, ed. *Cross-national research in sociology.* Newbury Park: Sage.

Limonard, S. & de Koning, N., 2005. Dealing with dilemmas in pre-competitive ICT development projects: the construction of 'the social' in designing new technologies. In L. Haddon, et al., eds. *Everyday innovators: researching the role of users in shaping ICTs.* Dordrect: Springer, pp.168-183.

Livingstone, S., 2003. On the challenges of cross-national comparative media research. *European Journal of Communication,* 18 (4) 477-500.

Livingstone, S. & Bovill, M. eds. 2001. *Children and their changing media environment. A European comparative study.* New Jersey: Lawrence Erlbaum Associates, Inc.

Livingstone, S., & Haddon, L., eds. 2009. *Kids online. Opportunities and risks for children.* Bristol: Policy Press.

Lobe, B., Livingstone, L & Haddon, L., 2007. *Researching children's experiences online across countries: issues and problems in methodology.* [Online] Available at: http://www.lse.ac.uk/collections/EUKidsOnline/Reports/Default.htm [Accessed 29 December 2009].

Lobe, B. Livingstone, S. Ólafsson, K. & Simões, J, eds. 2008. *Best practice research: how to research children and online technologies in comparative perspective.* [Online] Available at: http://www.lse.ac.uk/collections/EUKidsOnline/Reports/Default.htm [Accessed 29 December 2009].

Lobe, B. Simões, J.A. & Zaman, B., 2009. Research with children. In: S. Livingstone & L. Haddon, eds. *Kids online. Opportunities and risks for children.* Bristol: Policy Press, pp.31-41.

Ponte, C. Bauwens, J. & Mascheroni, G., 2009. Reporting young people and the internet: agency, voices and agendas. In: S. Livingstone & L. Haddon, eds. *Kids online. Opportunities and risks for children.* Bristol: Policy Press, pp.159-172.

Roberts, H., ed. 1981. Doing feminist research. London: Routledge and Kegan Paul.

Staksrud, E. Livingstone, L & Haddon, L., 2007. *What do we know about children's use of online technologies? A report on data vvailability and research gaps in Europe.* [Online] Available at: http://www.lse.ac.uk/

collections/EUKidsOnline/Reports/Default.htm [Accessed 29 December 2009].
Stald, G. & Haddon, L., 2008. *Cross-cultural contexts of research: factors influencing the study of children and the internet in Europe*. [Online] Available at: http://www.lse.ac.uk/collections/EUKidsOnline/Reports/Default.htm [Accessed 29 December 2009].
Swanson, D., 1992. Managing theoretical diversity in cross-national studies of political communication. In: J. Blumer, J. McLeod & K. Rosengren, eds. *Comparatively speaking: communication and culture across space and time*. Newbury Park: Sage, pp.19-34.

Vesna Dolničar

Chapter ten. Key methodological fallacies of digital divide research

Introduction

Regardless of the theoretical approach, technologies considered, level and units of observation, etc., empirical studies of the digital divide are exposed to certain common methodological problems. This chapter covers some issues we came across while selecting, analysing and interpreting the Eurostat data that attempt to measure the digital divide (see chapter 6 in this book). It provides an overview of some of the key methodological fallacies of digital divide research, before focusing on the questioning of monitoring changes in the digital divide, analysing secondary data, over time. Before attempting to systematise the different methodological challenges raised by digital divide measurements, let us briefly conceptualise the notion of the digital divide itself because this has implications for how we interpret what we measure.

In some of the first discussions of the concept, the term 'digital divide' was predominantly defined by differences between people who have access to digital technologies and those who do not – a binary division between 'haves' and 'have-nots', which in principle we can attempt to measure with statistics. However, current discussions of the digital divide, and of social exclusion more generally, are in turn based on older debates about relative poverty and deprivation. In these debates, one reason for using the word 'relative' when we discuss concepts like 'deprivation' is that any disadvantages can themselves be partial. This 'means that any 'have-nots', those on the wrong side of the digital divide, are not disadvantaged in every aspect of their life simply because of their lack of access to particular information and communication technologies (ICTs). They may be disadvantaged in certain respects, but not in others' (Haddon, 2004, p.21). Looking at statistics of have-nots hides this reality and can lead us to overestimate how disadvantaged they really are.

A similar point can be made when looking at statistics showing the changing of the digital divide, the adoption of ICTs, over time. It is all too easy to see population segments or countries that accept ICTs at a later stage – so-called 'late majorities' and 'laggards', using the terminology of the Roger's (1995) diffusion of innovations theory – as being somehow inferior. On the contrary, Fortunati and Manganelli (2002, p.6) argue that 'each country, each community and each social group will always find its own path towards technology'.

Further, Fortunati and Manganelli (2002, p.6) wrote that 'each culture, each society, negotiates the quantity of technology that it needs, in the same way as it decides which technology it needs most, starting from the specific environmental, social and productive conditions in which it lives'. Similarly, Menou and Taylor (2006) emphasised that without proper ethnographic knowledge of the population under study the use of any standard statistical measure to capture the digital divide might be grossly misleading. In line with this, we do not imply that all countries for example should strive for the highest level of ICT adoption.

Then there are issues of definition itself, which have implications for what we measure (and interpreting what we measure). The digital divide discussions are characterised by inconsistencies in use of the concept and they lack a uniform definition – which may in part hindered by the swift changes seen in the field of ICT. For example, in one of the earlier statements, the OECD (2001) described the digital divide as differences between individuals, households, companies and regions related to access to and usage of information and communication technologies (ICTs). This might be consider the basic (or so-called 'first') digital divide. However, later writers have pointed to other inequalities, noting the digital divide has multiple dimensions. Hence, any simple differentiation between those who have access to and use new ICTs and those that do not is insufficient. In fact, several extensions or types of digital divide have been identified, usually referring to one of the following:
– differences between segments of the population that have access to and/or use ICTs and those that do not;
– differences (usually among non-users) in terms of (lack of) interest in ICTs or in terms of obstacles to using ICTs;
– differences among users in terms of the characteristics of ICTs (e.g. fast/slow access to the internet – as discussed in chapter five); and
– differences related to people's experiences, skills and modes of using ICTs.

These last two dimensions of the digital divide reflect the range of experiences of internet use and the varied dimensions of digital inequality, focusing on the quality of equipment, user skills and differences in types of uses (for a more in-depth discussion of this see, e.g., DiMaggio, et al., 2001; DiMaggio, et al., 2004; Hargittai & Hinnant, 2008; Mossberger, et al., 2003; Selwyn, 2004; van Dijk, 2005). In fact, chapter six outlined the spectrum of online activities and the so-called 'experience gap' (in terms of involvement in various activities online) in relation to access to broadband from home[83]. The point is, which statistics you

---

83 Although it is important to acknowledge these multiple dimensions, the next methodological section and certainly the section measuring the dynamics of the digital divide over time will be limited to cases of the basic, i.e. the first, digital divide. As such, it will be operationalised by examining differences between countries in terms of broadband internet access at home. This limitation is imposed for practical reasons.

choose to look at, based on these different understandings and on what is included in the definition, can give very different pictures of the digital divide.

Let us discuss some of the methodological problems and examples of different digital divide indicators or benchmarks (i.e. reference points for making comparisons).

Defining and operationalising indicators

One of the important decisions researchers must make when studying the digital divide is the choice of indicator, given that there are numerous internet-related indicators and each has its own methodological specificities. Moreover, different surveys choose different indicators, sometimes with slightly different wording, which is particular problematic for cross-country comparisons. It is only in the past few years that there have been more systematic efforts to standardise official statistics. Let us observe the key methodological problems associated with one of the most commonly used indicators for studying the basic digital divide, i.e. the indicators that measure internet use and users (also see Vehovar & Dolničar, 2004). The variation in particular definitions of internet usage and users leads to different, inconsistent estimates of the percentage of internet users. Even the slightest alteration in wording can dramatically change the results.

What counts as internet usage?

When asking about internet usage, most typically only the internet is mentioned in a survey question. But more and more often the definition also explicitly includes the use of e-mail. Here we instantly encounter the problem that any answers about email might include non-internet-based email systems. Yet other definitions include Wireless Application Protocol. Some questionnaires thus have separate questions for web, e-mail, and other services, while others enquire about general internet usage. Overall, no common international standards have yet been accepted as regards what to include or what wording to use. In fact, any attempt to establish long-term conceptual guidelines for measuring and standardising internet-related indicators may be jeopardised by the (un-predictable) appearance of yet more devices that will enable access to the internet and by the evolution of innovative web-based services, especially those based on broadband. The term broadband is commonly used to describe 'always-on' internet connections that are significantly faster than dial-up technologies, but it does not refer to a specific speed or service. Currently 256 kbit/s is taken as the minimum speed. Recommendation I.113 of the ITU Standardisation Sector from 2003 defines broadband as a transmission capacity that is faster than the primary rate ISDN, at 1.5 or 2.0 Mbit/s. Elsewhere, broadband is considered to correspond to transmission speeds equal to or greater than 256 kbit/s, and some operators even label the basic rate ISDN (at 144 kbit/s) as a 'type of broadband' A review of contemporary studies shows that most often speed rates are considered: up to 512 kbps, 512 kbps-1 Mbps, 1-2 Mbps, 2-8 Mbps, 8-20 Mbps, > 20 Mbps, where 1-2 and 2-8 Mbps seem to be most useful to apply when searching for differences between and within EU countries. Another important characteristic of broadband is the capacity to share network access across several computers (Ewing & Thomas, 2008). In addition, the advent of mobile broadband, e.g. third generation mobile systems (3G) cannot be ignored. The ITU's (2006) report digital.life states that at the end of 2005 there were some 62 million mobile broadband users, with services launched in around 60 economies. Wireless local and metropolitan area networks (e.g. Wi-Fi, WiMax) also add to the picture, as does the development of ubiquitous networks and pervasive computing based on technologies like RFID (Radio Frequency Identification) and sensor-based networks. The implied question is that faced by a multiplicity of elements such as these, what counts as 'using the internet' when we start to think about the statistics that may measure the internet in the long term?

Time frames

There are also many different ways to operationalise the time frame considered in questions related to internet usage. Usage during the last three months is one common time span. Yet even more often the internet user is defined with a simple self-classification where the time frame is not even specified. Yet we know that the choice of time frame specified in questions makes a difference to the answers in the statistics. If you ask 'Do you use the internet at least once a month' this produces a 3-5% greater positive response compared to asking people if they currently use the internet. Moreover, asking about usage during the past three months produces at least 30% more users compared to asking if people are monthly users (Vehovar & Dolničar, 2004). The figures shrink even further when asking whether people are weekly or daily users.

Another standard way to measure ICT usage is to measure the time spent using these technologies. However, measures of duration are also problematic and so it is worth illustrating why this is so before looking at any figures. If we think of individual internet user there might be various reasons why someone spends many hours 'using' the service. For example, a service might be designed (primarily) for broadband users, but someone tries to download material from this service using a narrowband channel. Expanded to the country level, a country with less broadband overall might record more hours of usage if such downloading is widespread. In other words, the question of internet speed should be considered when collecting data about the time spent by users online. In addition, it should be pointed out that being continuously connected does not mean that one is using the internet all the time. In the case of broadband, 'always on' connections, it is particularly difficult to provide an estimation of the average time using the internet and to distinguish between time *spent* online (i.e. when the connection exists) and time *using* the internet (when someone is actively doing something online)[84]. Therefore, when phrasing survey questions it is very important to specify clearly the distinction between these two possibilities. If this is not done (and it is usually not) the validity of data is questionable because of measurement error. And when interpreting results showing time spent online we should thus consider the implications of the flat-rate tariffs (associated with broadband) which allow one to be 'always on'.

In the future, the definitions may have to become much more complex so the potential danger of creating improper comparisons will rise. While the development of standardised survey questions is thus extremely important, one has to bear in mind, as implied in the last section, that indicators will inevitably have to be redefined on a continuous basis as the internet evolves. Of course, these changes in operational definitions should reflect changes in the conceptualisation of the internet.

Structure of the targeted population

The number of internet users is often calculated as a share of the total population of a country, which acts as the base and a denominator for calculating percentages. This might lead to quite unfair comparisons because of national populations' varying age structures and may produce artificially low figures for certain countries (e.g. if they have a large proportion of very young children). Partly to address this, often only those aged category 18+ are included in research, particularly in the United States. In Europe, users older than 15 years have become the standard population. The Eurostat surveys include all individuals aged 16 to 74. The population aged 15 to 65 is also sometimes used as a basis for calculations, whereas media studies usually target the population aged 12 to 65 or 10 to 75. For a country with (broadband) internet penetration

---

84 For a more detailed discussion of this see the COST 298 interim report on the broadband digital divide; available at: http://www.cost298.org/uploadi/editor/1241179473BBDigital Divide_interim_report.pdf. [Accessed 27 January 2010]

reaching about a quarter of the total population, discrepancies arise from choosing the upper and lower age limits of the national populations (i.e. providing the basis for the denominator) ranging from the lowest internet penetration rate of 20% of the population aged 15+ to the highest penetration rate of 30% in a population aged 15 to 65.

Researchers making international comparisons of ICT adoption are thus forced to constantly search for and renegotiate the balance between standardisation and cross-country comparability (e.g. taking into account population age structures) and adapting to the rapid technological changes (discussed earlier).

When studying specifically the digital divide over time, we usually refer to simple percentage differences in internet usage across different segments. In order to overcome the problematic one-dimensional presentation of digital divide indicators, the next section suggests a new way of presenting results based on the time-distance methodology.

Monitoring changes in the digital divide in time

In the last few decades, developed countries have been characterised by a large increase in the adoption of new ICTs. Norris (2001: 26), one of the leading researchers of the digital divide in modern societies, provides a vivid description of digital divide studies as 'blurred snapshots of a moving bullet' (due to the incessant and rapid changes). Therefore, measuring the dynamics of the digital divide is one of the most important challenges facing information society studies. Currently, most of the existing comparisons of ICT take-up among various countries and/or population segments are typically presented and interpreted with 'static' measures (e.g. percentage differences, ratios, Gini coefficients, the Theil index, coefficients of variation etc.). However, since comparisons also usually involve observing changes over time, using such benchmarks can turn out to be extremely problematic since straightforward comparisons of percentages may not suffice in a rapidly changing environment.

The key question when measuring the dynamics of the digital divide is how to establish whether the digital divide is expanding, shrinking or stagnating. But when calculating the differences between various population segments different statistical measures can give partial (and often contradictory) pictures of the size of the digital divide. This often leads to a situation that at best only allows a partial answer to that key question. In order to measure the dynamics of the digital divide, the simultaneous three-dimensional monitoring of static absolute and relative differences (to be illustrated below) along with a dynamic, innovative S-time-distance is needed.

The S-time-distance methodology was developed – at conceptual and applied[85] levels – by Sicherl (see, for example, 1978, 2003). In brief, 'time distance' generally means the difference in time when two events occurred (e.g. the difference in years or months between two compared groups that achieved a certain percentage of internet penetration). More specifically, the statistical measure 'S-time-distance' measures the distance in time that it took for the two series being compared to reach a specified level of indicator X (e.g. the difference in the time that it took for one country to reach a point where 50% of households have a certain ICT compared to the time it took for another country to reach that level). The observed distance in time (the number of years, quarters, months, etc.) is used as a temporal measure of the disparity between the two series (in the example, the disparity between the two countries). This way of measuring is intuitively understandable and can be usefully applied as an important analytical and presentation tool at various levels to a wide variety of fields[86]. Next, let us illustrate a simplified case that clearly demonstrates how three alternative statistical measures can lead to different conclusions about the dynamics of the digital divide.

Absolute differences, relative differences and S-time-distance

When comparing the adoption of ICTs over time between two or more countries, regions groups within the population, etc., two possibilities arise:
a) Variable values can be compared to measure the adoption of ICTs at given points in time (e.g. in 2009, one country's internet adoption was x% while another country's was y%). The digital divide is thus calculated – for each measured point in time (in the example, 2009) – based on the absolute or relative difference between values of the monitored variables.
b) Points in time can be compared at given variable values measuring the adoption of ICT (as was described in the previous section, asking how many years' difference is there between countries reaching a certain level of adoption). The digital divide – for every level of chosen variable (for example, 50% of the households have adopted a certain ICT) – is calculated on the basis of the difference between two time points.

---

85 This methodology is an approach that is increasingly used in the field of social sciences because it allows an additional understanding of complex time processes. This generic idea can be used with numerous additional applications; for example, Granger and Jeon (1997, 2003) further elaborated this approach for evaluating forecasting models.
86 For a more detailed presentation (at conceptual and applied levels) and an in-depth discussion of the time-distance methodology, consult, for example Sicherl, 2003, 2004, 2005, 2006, 2007.

In order to obtain a holistic insight into the dynamics of the changes involved in the adoption of new technologies within the groups being studied, the two above mentioned approaches will be compared: conventional statistical (absolute and relative) measures of the digital divide vs. S-time-distance. This will first be illustrated by using these three measures in a hypothetical case monitoring the dynamics of change in the adoption of an ICT in two groups (A and B) in the period from 2003 to 2006 (Figure 1). Group A is characterised by an ICT increase from 10% to 20%, and group B from 5% to 15%.

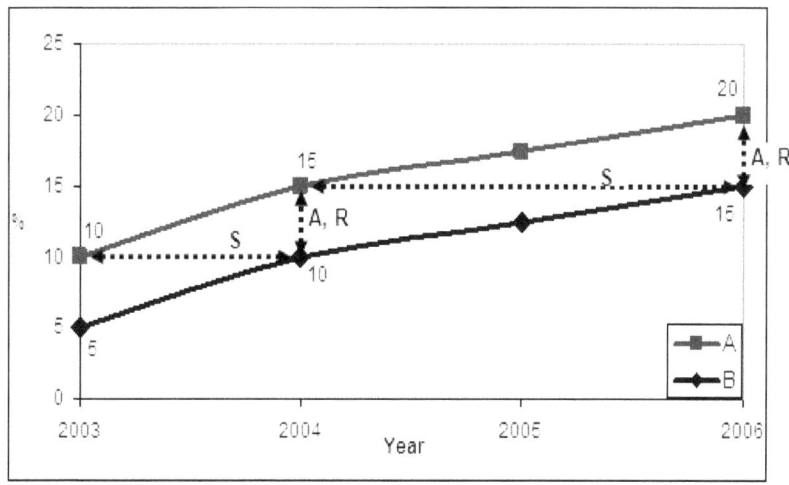

Figure 1: An example of the adoption of ICT in two compared groups (A and B) in the 2003 to 2006 period.

Was the digital divide growing or shrinking in the monitored time period? The calculation of the statistical measures and corresponding interpretations are as follows:
– Absolute differences are the most often used measures but they are also the most disputed. The absolute difference between groups A and B in 2003, is 5% (i.e. 10% for A minus 5% for B). In 2004 it is still 5% (15%-10%) and the same is true in 2006 (20%-15%). In other words, the absolute difference never changed over the years in this graph. But the earlier the year, (arguably) the more striking was the gap e.g. in 2003, A was double B – group B would have been seen to be really lagging behind group A in relative terms.

197

- Relative differences can be calculated in many ways (although this has no influence on the content aspect of the interpretation of differences). More usually when calculating the size of the digital divide between the studied groups, the ratio and percentage difference are used. The value of the ratio between groups A and B in 2003 was 0.5 (5 /10 =0.5); in 2004 it was 0.67 (10 / 15 = 0.67); while in 2006 it was 0.75 (15 / 20 = 0.75). Taking these ratios into account, the digital divide is shrinking because in 2003 group B only had 50% of the number of internet users in group A; in 2004, group B climbed to 67% of the number of internet users in group A; and in 2005 it reached 75%. A slightly different measure, percentage differences show the same trend:
  - in 2003: –50% ((5 – 10) /10) × 100 = –50%;
  - in 2004: –33.3% ((10 – 15) / 15) × 100 = –33.3%; and
  - in 2006: –25% ((15 – 20) / 20) × 100 = –25%.
- According to this way of looking at the picture, in 2003, among individuals in group B, 50% adopted ICT compared to group A; in 2004 the relative situation of group B improved by nearly 20 percentage points as group B lagged behind group A by only 33.3%. In 2006 group B further improved its situation; in the adoption of ICT group B did 25% worse than group A. While in absolute terms differences between groups A and B persist, relatively speaking the divide is shrinking over time.
- S-time-distance expresses the time (here, years) within which the lagging group B will achieve the level that group A is at today. Thus, in 2004 group B lags behind by one year because, B achieved group A's 2003 level of 10% with a one year delay(see the left horizontal dotted line in Figure 1). In 2006, the difference of 15% (B) and 20% (A) indicates that B is now lagging two years behind group A: it took two years from 2004-2006 for B to reach A's 15% ICT penetration level (see the right horizontal dotted line in Figure 1) The time delay of group B behind group A is therefore increasing; in terms of time, this indicates a growing digital divide.

The above illustration demonstrates that a simple example can yield three completely different results, moreover with different directions of change regarding the growth or decrease of the digital divide:
- regarding absolute differences the divide is stagnating;
- regarding relative differences the divide is shrinking; and
- regarding S-time-distance the divide is growing.

In order to define more accurately the relationships between absolute differences, ratios and S-time-distance, we can consider a so-called 'standard typology of relationships' among the statistical measures of the digital divide. Since each of the three measures may have three outcomes (increase, decrease or remain constant) the direction of change of absolute and relative statistical measures, and S-time-distance can be combined in 27 ways. This shows the potential relationships between the three measures.

Comparing Britain and Cyprus with the different measures

An illustrative example of the considerable differences that can exist regarding all three statistical measures is presented in Figure 2, which depicts the take-up of internet connections in British and Cypriot households in the period between 2003 and 2008. These two countries were chosen based on the secondary data analysis presented in a chapter six, where five countries were compared in terms of their broadband penetration rates. While the UK is the leading country in terms of broadband adoption among the five observed counties, Cyprus is characterised by the lowest broadband penetration.
- The divide between the United Kingdom and Cyprus has been constantly decreasing in terms of relative differences (in 2004 Cyprus was only 13% of the UK level, whereas in 2008 Cyprus reached 53% of the UK level).
- For most years the divide has been increasing with regard to absolute differences (with the exception of the last year, i.e. 2008, in which absolute differences started to decrease).

Absolute and relative differences thus yield contradictory conclusions, with the exception of the last year (2008), in which the gap is narrowing in terms of both statistical measures. However, even in this situation we should warn against oversimplified findings based on merely static measures.
- The corresponding time lag increased by 2008 (being 36 months in 2008, meaning that the rate of broadband penetration in Cyprus in 2008 had already been reached in the UK three years earlier)[87].

---

[87] The values of the three statistical measures reflect a number of factors. One is the shape of the diffusion functions being compared (where – in the case of the diffusion of ICTs – an increasing growth trend is often followed by a decreasing growth trend). Another is the initial time delay in the early stage of ICT adoption. A third is the final level of ICT penetration. Thus, when forecasting future trends of ICT penetration an explicit consideration of how the measures work in scenarios such as those in figures 1 and 2 is extremely useful for understanding the past and future dynamics of the digital divide. For a more in-depth discussion of these issues, see Dolničar (2008).

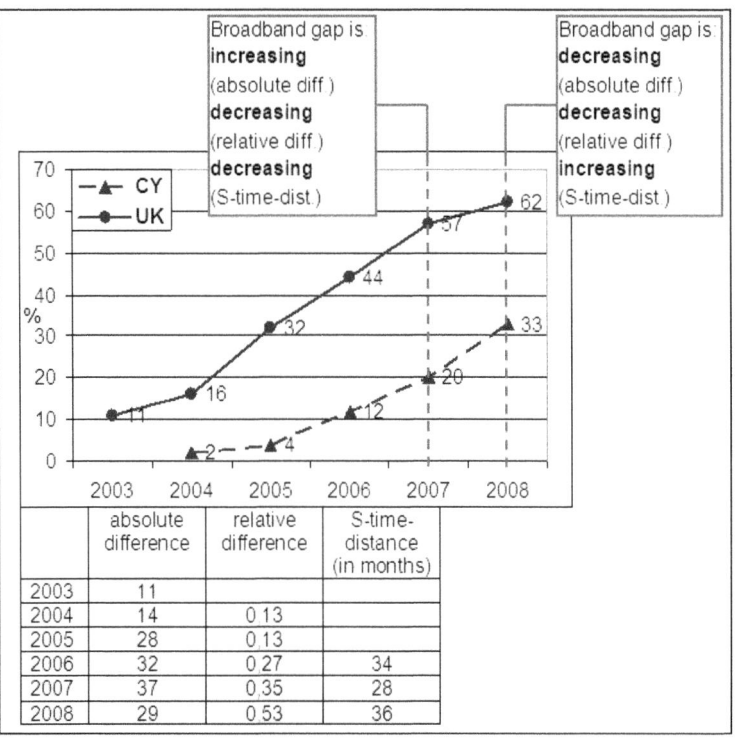

Figure 2: Comparison of three different statistical measures when computing differences between households with a broadband connection in Cyprus and in the United Kingdom. Source of data: Eurostat

Due to the complex nature of the relationship between the values of these measures, it is advisable to present them one after another, particularly if they lead to different evaluations of the process taking place. Nevertheless, providing all three measures offers some protection against those who object that only supplying one of these measures is producing misleading estimates about the size of differences in the adoption of ICTs.

Although reporting the three statistical measures is more informative (and holistic), because each of these measures may lead to different conclusions this may have the effect of leading to an almost complete relativisation of the issue being discussed (in the case the digital divide) because the answer becomes 'it all depends how you look at it'. The (at first) seemingly simple question about whether the digital divide is growing or shrinking cannot (in most ICTs and over most time periods) be answered unequivocally. From the statistical point of view, the above example yields an entirely unclear interpretation of whether the digital divide is growing or not.

This possible methodological quandary, an utterly blurred field that leaves no uniform answer to the key question about whether the digital divide is growing or shrinking, may be overcome in various ways. One is by developing an integral statistical measure which would be 'a weighted combination of static and time distance measures', i.e. a combination of the three measures (Sicherl, 2006, p.10). However, this leads to a series of additional questions. On what basis should the weights of each of the three statistical measures be defined (i.e. when combing them, should some be given more prominence)? Should these three measures always be given equal weights when combined, in all cases? Or should more weight (more prominence in the calculations) be assigned to the measure that would show largest differences between the compared groups? Should the weighting procedure depend on the issue being studied (i.e. a particular digital divide) and/or growth rate? How much should any weighting be influenced by the subjective evaluation of the different measures by the researchers involved (and do those interested in these figures regard the time dimension as being more or less useful to know than static dimensions of disparities)? How should we interpret the result if the sum of the three statistical measures is close to zero (that is, although each of the measures displays different trends, when combined the sum is zero): in this case, should we conclude that the differences are neither diminishing nor increasing?

So while this section of the chapter has tried to show how our understanding of dynamics can be enhanced by considering all three measures, this then raises new issues for researchers. Here we see examples of the next set of questions that emerge and pose new challenges for further research and demand a search for possibly more appropriate analytical solutions to studying the dynamics of the digital divide.

Conclusion

Even though there has been a proliferation of digital divide studies and corresponding indicators in recent years, these are still 'far from responsive to the needs of many stakeholders' (Menou & Taylor, 2006, p.261) and involve serious limitations. When systematising some of the methodological challenges faced in measuring digital divides, this chapter discussed issues in two areas. The first area related to variations and changes in the definition of what counts as internet usage together with how the indicators were operationalised. The second involved comparing results arising from the three different statistical measures that can be used when interpreting the dynamics of the digital divide.

As regards the first area, there are clearly different possible operational definitions of internet usage related to the various web-based services that can be explicitly mentioned in the question(s) and to various devices that enable access to the internet. Next, in order to illustrate some of the limitations of alternative ways of operationalising measurements the chapter focused on changes in wording that result in very different estimates of percentage shares of internet users. In particular, the time frame referred to when measuring internet usage can be very different (from one day to three months or even a year) or can even be disregarded, i.e. not specified in the question. This again influences the over- or underestimation of the number of internet users. The chapter also showed how differences in the structure of the targeted population can result in discrepancies between different estimates. Internet penetration may vary dramatically (a range of 20-30%), from being a lower share of the population defined as being aged 15+ to a higher share if we consider the population aged 15-65.

The other objective of the chapter was to demonstrate that an investigation of the digital divide in the temporal perspective requires the presentation of all three statistical measures (absolute differences, relative differences and S-time-distance)[88]. The key concern here is to establish whether the digital divide is expanding, shrinking or stagnating. Any indicator that relies solely on comparisons of absolute or of relative achievements may not be exhaustive in explaining the phenomena. Researchers should carefully take the time dimension into consideration, since the degree of disparity may be very different when comparing a static measure with a time-based one. If this requirement is not met, the results could prove to be biased and misleading. If one does not explicitly use the broader framework outlined here, the possibility arises that in political debate and policy formulation 'various interest groups would intentionally look only at the measure which will suit their particular interest' (Sicherl, 2007, p.241).

On one hand, there are many comparable empirically focused surveys that routinely monitor cross-national differences (in percentage terms) either by means of simple standardised benchmarks or by more complex indices. In this case, differences in access, (non-)usage and type of ICTs used are usually taken into account. On the other hand, albeit less frequently, some researchers are conducting more theoretically-based studies where the characteristics of (non-) users are also being examined. One may anticipate that we will see an increase in research efforts involving the latter approach to studying digital divide

---

88 The proposed framework for measuring the digital divide over time can be potentially extended to analyse other elements of digital divides (where motivation, willingness and skills for using new ICTs are studied; for a further discussion see, e.g. Vehovar, et al., 2006; Haddon, 2004) and other factors, ICTs, units and levels of study.

phenomena in the future. Given that the penetration rate of the basic ICT infrastructure is growing rapidly (with the exception of some regions), the next questions of why ICTs are adopted and used, how this is taking place, and what benefits follow for the individuals are using ICTs will have to be addressed more often. Nevertheless, the significance of the need to establish proper measurement procedures for even the most basic internet-related statistics should not be underestimated. This is because such statistics often form the basis for further comparisons, and open the way for a more in-depth understanding of the digital divide concept. As shown in this chapter, even when measuring the so-called basic (or first) digital divide, many components related to measurement, presentation and interpretation of the data can dramatically influence our understanding of the digital divide. Thus, the methodological issues (on the levels of operationalisation and analysis) presented in this chapter cannot be ignored when evaluating digital divide phenomena.

Bibliography

DiMaggio, P. Hargittai, E. Neuman, W. R. & Robinson, J. P., 2001. Social implications of the Internet. *Annual Review of Sociology*, 27, pp.307-336.
DiMaggio, P. Hargittai, E. Celeste, C. & Shafer, S., 2004. Digital inequality: from unequal access to differentiated use. In: K. Neckerman, ed. *Social Inequality*. New York: Russel Sage Foundation, pp.355-400.
Dolničar, V., 2008. *Merjenje dinamike digitalnega razkoraka [Measuring the dynamics of the digital divide]*. Ljubljana: Znanstvena knjižnica.
Ewing, S. & Thomas, J., 2008. Broadband and the 'creative internet': Australians as consumers and producers of cultural content online. *Observatorio (OBS*) Journal*, 2, pp.187-208. [Online] Available at: http://www.obs.obercom.pt/index.php/obs/article/view/215/190 [Accessed 14 December 2009].
Fortunati, L. & Manganelli, A., 2002. A review on the literature on gender and ICTs in Italy. In: K.H. Sørensen & J. Stewart, eds. *Digital divides and inclusion measures. A review of literature and statistical trends on gender and ICT*. Trondheim/Edinburgh: NTNU, pp. 137-170.
Granger, C.W.J. & Jeon, Y., 1997. Measuring lag structure in forecasting models - the introduction of time distance. *Economics Working Paper Series 97–24*. San Diego: University of California, Department of Economics.
Granger, C.W.J. & Jeon, Y., 2003. A time-distance criterion for evaluating forecasting models. *International Journal of Forecasting*, 19 (2), pp.199-215.

Haddon, L., 2004. *Information and communication technologies in everyday life: A concise introduction and research guide*. Oxford: Berg.

Hargittai, E. & Hinnant, A., 2008. Digital inequality: differences in young adults' use of the internet. *Communication Research*, 35(5), pp.602-621.

ITU, 2006. *Digital.life: ITU internet report 2006.* Geneva: International Telecommunication Union (ITU). [Online] Available at: http://www.itu.int/osg/spu/publications/digitalife/docs/digital-life-web.pdf [Accessed 14 December 2009]

Menou, M.J. & Taylor, R.D., 2006. A "grand challenge": Measuring information societies. *The Information Society*, 22(5), pp.261-267.

Mossberger, K. Tolbert, C.J. & Stansbury, M., 2003. *Virtual inequality: beyond the digital divide*. Washington, DC: Georgetown University Press.

Norris, P., 2001. *Digital divide: civic engagement, information poverty, and the Internet worldwide*. Cambridge, New York: Cambridge University Press.

OECD, 2001. *Understanding the digital divide*. Paris: OECD Publications.

Rogers, E., 1995. *Diffusion of innovations.* New York: The Free Press.

Selwyn, N., 2004. Reconsidering political and popular understandings of the digital divide. *New Media and Society*, 6(3), pp.341-362.

Sicherl, P., 1978. S-distance as a measure of time dimension of disparities. In: Z. Mlinar & H. Teune, eds. *The Social Ecology of Change*. London and Beverly Hills: Sage Publications.

Sicherl, P., 2003. Different statistical measures provide different perspectives on digital divide *6th Conference of the European Sociological Association*. Murcia, Spain 23-27 September 2003. [Online] Available at: http://www.sicenter.si/pub/Sicherl_Digital_divide_Murcia.pdf [Accessed 14 December 2009].

Sicherl, P., 2004. *A new generic statistical measure in dynamic gap analysis. The European e-Business report*. Luxembourg: European Commission.

Sicherl, P., 2005. Analysis of information society indicators with time distance methodology. *Journal of Computing and Information Technology*, 13(4), pp.293-298.

Sicherl, P., 2006. The intertemporal aspects of well-being and societal progress. *Joint OECD-JRC workshop Measuring Well-being and Societal Progress*. Milan, Italy 19-21 June 2006.

Sicherl, P., 2007. The inter-temporal aspect of well-being and societal progress. *Social Indicators Research*, 84(2), pp.231-247.

van Dijk, J.A.G.M., 2005. *The deepening divide: inequality in the information society*. Thousand Oaks: Sage.

Vehovar, V. & Dolničar, V., 2004. Benchmarking internet. In: H. Bidgoli, ed. *The Internet Encyclopaedia*. Hoboken, NJ: John Wiley & Sons, pp.57-71.

Vehovar, V. Sicherl P. Hüsing T. & Dolnicar V., 2006. Methodological challenges of digital divide measurements. *The Information Society*, 22(5), pp.279-290.

Authors

Annika Bergström, PhD, is researcher and senior lecturer in the Media and Communication Science at the Department of Journalism, Media and Communication (JMG), University of Gothenburg. She holds a Ph.D degree in Journalism and Mass Communication from the University of Gothenburg. Her dissertation outlines internet news habits from 1998-2003. Annika has been a part of The Newspaper Research Programme since 1997. Over the last five years her research has mainly focused on the use of web 2.0 applications, audience participation in journalism and citizen's political participation online.

Charalambos Christou is currently an Associate Professor in the Computer Science Department at the University of Nicosia, Cyprus. He is a member of the University's Senate. Dr Christou earned his BSc, MSc and Ph.D in Computer Engineering from the New Jersey Institute of Technology, USA. He worked in the USA as an IT Consultant, as a Visiting Professor at NJIT and as a Lecturer at Essex County College. He conducted extensive work in the area of E-learning and Parallel Computers, and participated as a researcher in several projects funded by the EU and the Research Promotion Foundation of Cyprus.

Vesna Dolničar, PhD, is a teaching assistant in the field of social informatics and methodology and a researcher at the Centre for Methodology and Informatics, Faculty of Social Sciences, University of Ljubljana. She lectures on several methodological courses and has been actively involved in (inter)national research projects related to the field of measuring, monitoring and understanding the information society and the specific needs and motives of potentially excluded groups. She is currently a national correspondent for the 6th Framework Programme project SOPRANO.

Pedro Gómez Fernández, Communication and Journalism Studies Scholar and journalist, has taught Public Opinion and Political Communication at the University Complutense of Madrid. His research interest lies in the role of new technologies in political communication and database journalism. As a journalist he directed the social and cultural programmes for emigrants for the Spanish Broadcasting Overseas Service and directed the project introducing information systems and technology into Spanish Public Radio and TV (RTVE) for newsroom services.

Rosemarie Gannon has a PhD from University College Dublin. Her PhD explored the spatial dimension of an information society and investigated the friction of distance in a rural Irish community. Her research interests include the social and cultural impacts of ICTs, ICT use (non-use) in everyday life and the adoption and use of ICTs in rural areas. Previously Rosemarie worked as a project manager/researcher at the European Institute for the Media in Düsseldorf.

Fruzsina Gyenes is a cultural anthropologist and researcher at the Excenter Researcher Foundation. She worked as a researcher in the Information Society Research Institute (2007-2009) and in Apáczai Foundation (2009-2010). She participated in many research programmes in various fields including Hungarian minorities in abroad, youth research, educational research and research on the information society. She was one of the editors of the Hungarian Information Society Quarterly (2007-2008).

Leslie Haddon, PhD, is Senior Researcher and part-time Lecturer in the Department of Media and Communications at the LSE. His work focuses on the social shaping and consumption of ICT. Recent works include Haddon, L. (2004) *Information and Communication Technologies in Everyday Life: A Concise Introduction and Research Guide,* Berg; Haddon, L, et al (eds) (2005) *Everyday Innovators, Researching the Role of Users in Shaping ICTs*, Springer; Loos, E., Haddon, L. and Mante-Meijer, E. (eds) (2008) *The Social Dynamics of Information and Communications Technology,* Ashgate; Livingstone, S., and Haddon, L. (eds) (2009) *Kids Online. Opportunities and Risks for Children,* Policy Press; Green, N. and Haddon, L. (2009) *Mobile Communications. An Introduction to New Media*, Berg, Oxford

Peter L. Heinzmann received the Dipl. El. Ing. and Dr. Sc. Techn. degrees from the Swiss Federal Institute of Technology, ETH Zurich, Switzerland, in 1979 and 1987, respectively. After research assignments at the Institute for Communication Technology, ETH Zurich and at the IBM Research Division, Zurich Research Laboratory, Rueschlikon, Switzerland he joined the faculty of the University of Applied Sciences in Rapperswil (HSR) in 1991, where he still teaches Computernetworks and Internet Security. In 1997 Prof. Heinzmann founded the cnlab AG, a company which develops internet applications like performance measurement systems, traffic monitors or webcams and weather stations.

Giovanna Mascheroni, PhD, is Research Fellow at the Department of Political Studies, University of Torino, where she teaches New Media and Communication. She also works as a senior researcher at OssCom (a media research centre based at Università Cattolica). Her current research interests are focused on network and mobile sociality, uses of social media and youth participation online. Among her international publications are Mascheroni, G. (2007), 'Global nomads' network and mobile sociality', *Information, Communication & Society*, 10(4), and Ponte, C., Bawens, J., Mascheroni, G. (2009) "Children and the internet in the news: agency, voices and agenda ", in Livingstone, S., Haddon, L. (Eds.) *Kids Online. Opportunities and Risks for Children*, The Policy Press, Bristol, pp. 159-171.

Robert Pinter, PhD, is the online strategic director of the Ipsos market research company in Hungary. Until 2008 he worked as scientific strategic director of the Information Society Research Institute and as assistant professor in the Information and Knowledge Management Department of Budapest University of Technology and Economics. He was the editor-in-chief of the Hungarian Information Society Quarterly (2005-2008). In the last decade he participated in many studies of information society in Hungary, covered topics like new media and internet use, internet diffusion, digital divide, e-government, information policy, and the ICT use of NGOs.

Pedro Puga is a sociologist, graduated from ISCTE, investigating the areas of arts, culture, social memory and simulacra. He holds a Master Degree in Public Administration from ISCTE-IUL. He has already participated in research projects on Youth and Solitude (ICS), Politics/Corruption (CIES-ISCTE), Young People in Portugal, and Health and ICT (CIES-ISCTE). He is a member of Cost 605 Econ@Tel - A Telecommunications Economics COST Network and worked as a researcher for OberCom (Observatory for the Media) from 2007 to 2009. He is now working as a Media Researcher for ERC (the Portuguese Media Regulator).

Barbara Scifo, PhD, is Assistant Professor of Sociology at the Università Cattolica of Milan, where she teaches Languages of Mass Communication. She is on the Board of OssCom (a media research centre, directed by Fausto Colombo) as scientific coordinator, and she also works as senior researcher. She is interested in the social shaping of ICTs, with a particular focus on mobile communication, the internet and digital television. Among her publications are *Crossmedia cultures* (edited with F. Pasquali and N. Vittadini, Milano, 2010) dealing with youth and digital media consumptions; *Culture mobili* (Milano 2005), reporting studies of mobile phone use among teenagers; *Consumare la rete* (edited with F. Pasquali, Milano 2002), reporting two years of field work on internet consumption. She has published essays in Italian and in international journals and collective volumes.

Anna Sfardini, PhD, is Assistant Researcher at the Department of Social and Political Studies, University of Milan. Her research interests are focused on media convergence processes, the practices of media consumption, the processes of mediatisation and the popularisation of political discourse. She has published many essays in journals and chapters in collective books. She has also published *MultiTV. The television experience in the age of convergence* (co-edited with Massimo Scaglioni, Roma; Carocci, 2008), *Reality Tv* (Unicopli, Milano, 2009) and *Politica Pop* (co-edited with Gianpietro Mazzoleni, Il Mulino, Bologna, 2009).

Matteo Stefanelli, Ph.D., is Senior Analyst at OssCom, a media research centre based at the Università Cattolica of Milano, where he is also a Research Fellow in the Faculty of Political Sciences. His research interests cover the social history of media and the intersections between new media and content industry, with a special focus on graphic cultures. He edited *Spazi sensibili. Pubblicità, comunicazione e ambiente urbano* (with P.Aroldi and F.Pasquali, V&P, 2006), and *Media+Generations: Summary Report* (V&P, 2009). He has also published in Italian and international books and academic journals, such as *Médiamorphoses*, *OBS*Journal*, *Neuvième Art* and *Bianco&Nero*. He is the Founding Editor of the Journal *SIGNs - Studies in Graphic Narratives*.

Frank Thomas is an independent social scientist in the social studies of information and communication technologies working with the telecommunications industry. Among other topics, his studies dealt with the appropriation and usage of the internet in everyday life, the links between social capital, quality of life and ICT usage as a partner of EU's SOCQUIT project, and the trends in mobile urban life. Frank Thomas holds a Ph.D. from the Max-Planck-Institute for the Study of Societies at Cologne, Germany, based on sa study of the German telephone system as a large technical system. He is member of the jury of the national Villes Internet challenge for French municipalities.

Jorge Vieira is a researcher at OberCom, the Portuguese Observatory for the Media. His main research interests are music, new media, art and youth cultures. He is a PhD student in the Sociology Department at ISCTE-IUL (Lisbon University Institute). He holds a Master in Communication, Culture and Information Technologies from ISCTE with a dissertation on 'music 2.0' focusing on new patterns of consumption, networking and digital lifestyles. He also graduated in Sociology with a dissertation on 'Street Art' focusing on the areas of art, culture, youth and urban studies.

Nicoletta Vittadini is researcher in the sociology of culture and communication at the Catholic University of Milan where she teaches the sociology of communication and new media theory. She is a senior researcher at OssCom (Research Center on Media and Communication) and vice-director of Almed (Postgraduate School in Media, Communication and Performing Arts). Among her publications are *Digitizing Tv. Theoretical issues and comparative studies across europe* (ed. with Fausto Colombo) Vita e Pensiero, Milano 2006; *Immigrati, media e integrazione* (ed.) Ikon 54/55 Angeli, Milano 2008; *Incipit digitale. L'avvio della Dtt in Italia tra discorsi, prodotti e consumo* (ed. with M.G. Fanchi) Comunicazioni Sociali 1/2008, Milano.

Participation in Broadband Society

Edited by Leopoldina Fortunati / Julian Gebhardt / Jane Vincent

This series publishes peer-reviewed monographs and edited volumes by internationally renowed scholars in the field of the 'social use of information and communication technologies (mass media included)', 'communication studies' and 'science and technology studies'. It provides an editorial space specifically dedicated to the collection of work that integrates new research regarding theoretical discourse, methodologies and studies from multiple disciplines such as sociology, anthropology, psychology, geography, linguistics, information science, engeneering and more.

The editors particularly welcome texts elaborating new theories, original methodological approaches and challenges to existing knowledge. Proposals aimed at scholars, professionals and operators working in the diverse field of participation in broadband society are invited from all disciplines.

Vol. 1 Leopoldina Fortunati / Jane Vincent / Julian Gebhardt / Andraz Petrovčič / Olga Vershinskaya (eds.): Interacting with Broadband Society. 2010.

Vol. 2 Julian Gebhardt / Hajo Greif / Lilia Raycheva / Claire Lobet-Maris / Amparo Lasen (eds.): Experiencing Broadband Society. 2010.

Vol. 3 Leslie Haddon (ed.): The Contemporary Internet. National and Cross-National Studies. 2011.

Vol. 4 Hajo Greif / Larissa Hjorth / Amparo Lasén / Claire Lobet-Maris (eds.): Cultures of Participation. Media Practices, Politics and Literacy. 2011.

Vol. 5 Fausto Colombo / Leopoldina Fortunati (eds.): Broadband Society and Generational Changes. 2011.

www.peterlang.de

Ulrike Bauernfeind

User Satisfaction with Personalised Internet Applications

Frankfurt am Main, Berlin, Bern, Bruxelles, New York, Oxford, Wien, 2008.
190 pp., num. tab., graph.
Forschungsergebnisse der Wirtschaftsuniversität Wien. Vol. 27
ISBN 978-3-631-57770-7 · pb. € 39.–*

The study focuses on user satisfaction with websites and personalised internet applications in particular. The abundance of information on the web is increasing more and more. Therefore, the significance of websites targeting the users' preferences, like personalised Internet applications, is rising.
The aim of this study was to find out which factors determine user satisfaction with personalised internet applications. Factors like the usefulness of the information or trust towards how personal information is handled were considered. A large-scale user survey evaluating three internet applications (from the travel, e-learning and real estate domains) was conducted. Expert opinions were collected to complement the results and provide insights from users' and experts' points of views.

Contents: Personalised internet applications · Human – computer interaction · Technology acceptance model · Website evaluation · User satisfaction · Development of a research model · Structural equation modelling · Expert interviews · Success factors

Frankfurt am Main · Berlin · Bern · Bruxelles · New York · Oxford · Wien
Distribution: Verlag Peter Lang AG
Moosstr. 1, CH-2542 Pieterlen
Telefax 0041(0)32/3761727

*The €-price includes German tax rate
Prices are subject to change without notice
Homepage http://www.peterlang.de

www.ingramcontent.com/pod-product-compliance
Ingram Content Group UK Ltd.
Pitfield, Milton Keynes, MK11 3LW, UK
UKHW021823140426
5217IPUK00004B/55